SOLVING VIBRATION
ANALYSIS PROBLEMS
USING MATLAB

SOLVING VIBRATION
ANALYSIS PROBLEMS
USING MATLAB

Rao V. Dukkipati, Ph.D., P.E.

Fellow of ASME and CSME
Professor and Chair
Graduate Program Director
Department of Mechanical Engineering
Fairfield University
Fairfield, Connecticut
USA

PUBLISHING FOR ONE WORLD

NEW AGE INTERNATIONAL (P) LIMITED, PUBLISHERS
(formerly Wiley Eastern Limited)
New Delhi • Bangalore • Chennai • Cochin • Guwahati • Hyderabad
Jalandhar • Kolkata • Lucknow • Mumbai • Ranchi
Visit us at **www.newagepublishers.com**

ISBN : 81-224-2064-8

Rs. 495.00

C-07-01-1251

Printed in India at Chaman Offset, Delhi.
Typeset at Goswami Associates, Delhi.

PUBLISHING FOR ONE WORLD

NEW AGE INTERNATIONAL (P) LIMITED, PUBLISHERS
(formerly Wiley Eastern Limited)
4835/24, Ansari Road, Daryaganj, New Delhi - 110002
Visit us at **www.newagepublishers.com**

To Lord Sri Venkateswara

Preface

Vibration Analysis is an exciting and challenging field and is a multidisciplinary subject. This book is designed and organized around the concepts of Vibration Analysis of Mechanical Systems as they have been developed for senior undergraduate course or graduate course for engineering students of all disciplines.

This book includes the coverage of classical methods of vibration analysis: matrix analysis, Laplace transforms and transfer functions. With this foundation of basic principles, the book provides opportunities to explore advanced topics in mechanical vibration analysis.

Chapter 1 presents a brief introduction to vibration analysis, and a review of the abstract concepts of analytical dynamics including the degrees of freedom, generalized coordinates, constraints, principle of virtual work and D'Alembert's principle for formulating the equations of motion for systems are introduced. Energy and momentum from both the Newtonian and analytical point of view are presented. The basic concepts and terminology used in mechanical vibration analysis, classification of vibration and elements of vibrating systems are discussed. The free vibration analysis of single degree of freedom of undamped translational and torsional systems, the concept of damping in mechanical systems, including viscous, structural, and Coulomb damping, the response to harmonic excitations are discussed. Chapter 1 also discusses the application such as systems with rotating eccentric masses; systems with harmonically moving support and vibration isolation ; and the response of a single degree of freedom system under general forcing functions are briefly introduced. Methods discussed include Fourier series, the convolution integral, Laplace transform, and numerical solution. The linear theory of free and forced vibration of two degree of freedom systems, matrix methods is introduced to study the multiple degrees of freedom systems. Coordinate coupling and principal coordinates, orthogonality of modes, and beat phenomenon are also discussed. The modal analysis procedure is used for the solution of forced vibration problems. A brief introduction to Lagrangian dynamics is presented. Using the concepts of generalized coordinates, principle of virtual work, and generalized forces, Lagrange's equations of motion are then derived for single and multi degree of freedom systems in terms of scalar energy and work quantities.

An introduction to MATLAB basics is presented in Chapter 2. Chapter 2 also presents MATLAB commands. MATLAB is considered as the software of choice. MATLAB can be used interactively and has an inventory of routines, called as functions, which minimize the task of programming even more. Further information on MATLAB can be obtained from: The MathWorks, Inc., 3 Apple Hill Drive, Natick, MA 01760. In the computational aspects, MATLAB has emerged as a very powerful tool for numerical computations involved in control systems engineering. The idea of computer-aided design and analysis using MATLAB with the Symbolic Math Tool Box, and the Control System Tool Box has been incorporated.

Chapter 3 consists of many solved problems that demonstrate the application of MATLAB to the vibration analysis of mechanical systems. Presentations are limited to linear vibrating systems.

Chapters 2 and 3 include a great number of worked examples and unsolved exercise problems to guide the student to understand the basic principles, concepts in vibration analysis engineering using MATLAB.

I sincerely hope that the final outcome of this book helps the students in developing an appreciation for the topic of engineering vibration analysis using MATLAB.

An extensive bibliography to guide the student to further sources of information on vibration analysis is provided at the end of the book. All end-of-chapter problems are fully solved in the Solution Manual available only to Instructors.

—Author

Acknowledgements

I am grateful to all those who have had a direct impact on this work. Many people working in the general areas of engineering system dynamics have influenced the format of this book. I would also like to thank and recognize undergraduate and graduate students in mechanical engineering program at Fairfield University over the years with whom I had the good fortune to teach and work and who contributed in some ways and provided feedback to the development of the material of this book. In addition, I am greatly indebted to all the authors of the articles listed in the bibliography of this book. Finally, I would very much like to acknowledge the encouragement, patience, and support provided by my wife, Sudha, and family members, Ravi, Madhavi, Anand, Ashwin, Raghav, and Vishwa who have also shared in all the pain, frustration, and fun of producing a manuscript.

I would appreciate being informed of errors, or receiving other comments and suggestions about the book. Please write to the author's Fairfield University address or send e-mail to Rdukkipati@mail.fairfield.edu.

Rao V. Dukkipati

Contents

CHAPTER **1**

Introduction to Mechanical Vibrations

Vibration is the motion of a particle or a body or system of connected bodies displaced from a position of equilibrium. Most vibrations are undesirable in machines and structures because they produce increased stresses, energy losses, cause added wear, increase bearing loads, induce fatigue, create passenger discomfort in vehicles, and absorb energy from the system. Rotating machine parts need careful balancing in order to prevent damage from vibrations.

Vibration occurs when a system is displaced from a position of stable equilibrium. The system tends to return to this equilibrium position under the action of restoring forces (such as the elastic forces, as for a mass attached to a spring, or gravitational forces, as for a simple pendulum). The system keeps moving back and forth across its position of equilibrium. A *system* is a combination of elements intended to act together to accomplish an objective. For example, an automobile is a system whose elements are the wheels, suspension, car body, and so forth. A *static* element is one whose output at any given time depends only on the input at that time while a *dynamic* element is one whose present output depends on past *inputs*. In the same way we also speak of *static* and *dynamic systems*. A *static system* contains all elements while a *dynamic system* contains at least one dynamic element.

A physical system undergoing a time-varying interchange or dissipation of energy among or within its elementary storage or dissipative devices is said to be in a *dynamic state*. All of the elements in general are called *passive*, *i.e.*, they are incapable of generating net energy. A dynamic system composed of a finite number of storage elements is said to be *lumped* or *discrete*, while a system containing elements, which are dense in physical space, is called *continuous*. The analytical description of the dynamics of the discrete case is a set of ordinary differential equations, while for the continuous case it is a set of partial differential equations. The analytical formation of a dynamic system depends upon the kinematic or geometric constraints and the physical laws governing the behaviour of the system.

1.1 CLASSIFICATION OF VIBRATIONS

Vibrations can be classified into three categories: *free, forced*, and *self-excited*. *Free vibration* of a system is vibration that occurs in the absence of external force. An external force that acts on the system causes forced vibrations. In this case, the exciting force continuously supplies energy to the system. Forced vibrations may be either deterministic or random (see Fig. 1.1). *Self-excited vibrations* are periodic and deterministic oscillations. Under certain conditions, the

1

equilibrium state in such a vibration system becomes unstable, and any disturbance causes the perturbations to grow until some effect limits any further growth. In contrast to forced vibrations, the exciting force is independent of the vibrations and can still persist even when the system is prevented from vibrating.

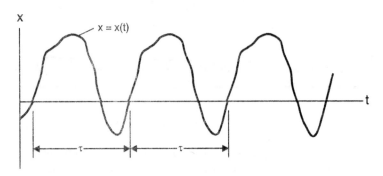

Fig. 1.1(a) A deterministic (periodic) excitation.

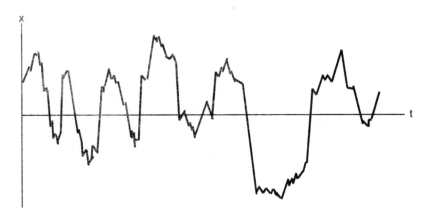

Fig. 1.1(b) Random excitation.

1.2 ELEMENTARY PARTS OF VIBRATING SYSTEMS

In general, a vibrating system consists of a spring (a means for storing potential energy), a mass or inertia (a means for storing kinetic energy), and a damper (a means by which energy is gradually lost) as shown in Fig. 1.2. An undamped vibrating system involves the transfer of its potential energy to kinetic energy and kinetic energy to potential energy, alternatively. In a damped vibrating system, some energy is dissipated in each cycle of vibration and should be replaced by an external source if a steady state of vibration is to be maintained.

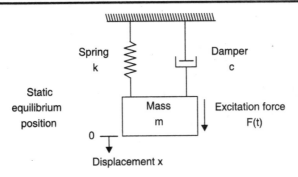

Fig. 1.2 Elementary parts of vibrating systems.

1.3 PERIODIC MOTION

When the motion is repeated in equal intervals of time, it is known as *periodic motion*. Simple harmonic motion is the simplest form of periodic motion. If $x(t)$ represents the displacement of a mass in a vibratory system, the motion can be expressed by the equation

$$x = A \cos \omega t = A \cos 2\pi \frac{t}{\tau}$$

where A is the amplitude of oscillation measured from the equilibrium position of the mass.

The repetition time τ is called the *period of the oscillation*, and its reciprocal, $f = \dfrac{1}{\tau}$, is called the *frequency*. Any periodic motion satisfies the relationship

$$x(t) = x(t + \tau)$$

That is   Period $\tau = \dfrac{2\pi}{\omega}$ s/cycle

Frequency $f = \dfrac{1}{\tau} = \dfrac{\omega}{2\pi}$ cycles/s, or Hz

ω is called the *circular frequency* measured in rad/sec.

The velocity and acceleration of a harmonic displacement are also harmonic of the same frequency, but lead the displacement by $\pi/2$ and π radians, respectively. When the acceleration $\ddot{X}$ of a particle with rectilinear motion is always proportional to its displacement from a fixed point on the path and is directed towards the fixed point, the particle is said to have *simple harmonic motion*.

The motion of many vibrating systems in general is not harmonic. In many cases the vibrations are periodic as in the impact force generated by a forging hammer. If $x(t)$ is a periodic function with period τ, its Fourier series representation is given by

$$x(t) = \frac{a_0}{2} + \sum_{n=1}^{\infty} (a_n \cos n\omega t + b_n \sin n\omega t)$$

where $\omega = 2\pi/\tau$ is the fundamental frequency and $a_0, a_1, a_2, ..., b_1, b_2, ...$ are constant coefficients, which are given by:

$$a_0 = \frac{2}{\tau} \int_0^{\tau} x(t)\, dt$$

$$a_n = \frac{2}{\tau} \int_0^\tau x(t) \cos n\omega t \, dt$$

$$b_n = \frac{2}{\tau} \int_0^\tau x(t) \sin n\omega t \, dt$$

The exponential form of $x(t)$ is given by:

$$x(t) = \sum_{n=-\infty}^{\infty} c_n e^{in\omega t}$$

The Fourier coefficients c_n can be determined, using

$$c_n = \frac{1}{\tau} \int_0^\tau (x)t \, e^{-in\omega t} \, dt$$

The harmonic functions $a_n \cos n\omega t$ or $b_n \sin n\omega t$ are known as the *harmonics of order n* of the periodic function $x(t)$. The harmonic of order n has a period τ/n. These harmonics can be plotted as vertical lines in a diagram of amplitude (a_n and b_n) versus frequency ($n\omega$) and is called *frequency spectrum*.

1.4 DISCRETE AND CONTINUOUS SYSTEMS

Most of the mechanical and structural systems can be described using a finite number of degrees of freedom. However, there are some systems, especially those include continuous elastic members, have an infinite number of degree of freedom. Most mechanical and structural systems have elastic (deformable) elements or components as members and hence have an infinite number of degrees of freedom. Systems which have a finite number of degrees of freedom are known as *discrete* or *lumped parameter systems*, and those systems with an infinite number of degrees of freedom are called *continuous* or *distributed systems*.

1.5 VIBRATION ANALYSIS

The outputs of a vibrating system, in general, depend upon the initial conditions, and external excitations. The vibration analysis of a physical system may be summarised by the four steps:

1. Mathematical Modelling of a Physical System
2. Formulation of Governing Equations
3. Mathematical Solution of the Governing Equations

1. Mathematical modelling of a physical system

The purpose of the mathematical modelling is to determine the existence and nature of the system, its features and aspects, and the physical elements or components involved in the physical system. Necessary assumptions are made to simplify the modelling. Implicit assumptions are used that include:

(a) A physical system can be treated as a continuous piece of matter

(b) Newton's laws of motion can be applied by assuming that the earth is an internal frame

(c) Ignore or neglect the relativistic effects

All components or elements of the physical system are linear. The resulting mathematical model may be linear or non-linear, depending on the given physical system. Generally speaking, all physical systems exhibit non-linear behaviour. Accurate mathematical modelling of any physical system will lead to non-linear differential equations governing the behaviour of the system. Often, these non-linear differential equations have either no solution or difficult to find a solution. Assumptions are made to linearise a system, which permits quick solutions for practical purposes. The advantages of linear models are the following:

(1) their response is proportional to input

(2) superposition is applicable

(3) they closely approximate the behaviour of many dynamic systems

(4) their response characteristics can be obtained from the form of system equations without a detailed solution

(5) a closed-form solution is often possible

(6) numerical analysis techniques are well developed, and

(7) they serve as a basis for understanding more complex non-linear system behaviours.

It should, however, be noted that in most non-linear problems it is not possible to obtain closed-form analytic solutions for the equations of motion. Therefore, a computer simulation is often used for the response analysis.

When analysing the results obtained from the mathematical model, one should realise that the mathematical model is only an approximation to the true or real physical system and therefore the actual behaviour of the system may be different.

2. Formulation of governing equations

Once the mathematical model is developed, we can apply the basic laws of nature and the principles of dynamics and obtain the differential equations that govern the behaviour of the system. A basic law of nature is a physical law that is applicable to all physical systems irrespective of the material from which the system is constructed. Different materials behave differently under different operating conditions. Constitutive equations provide information about the materials of which a system is made. Application of geometric constraints such as the kinematic relationship between displacement, velocity, and acceleration is often necessary to complete the mathematical modelling of the physical system. The application of geometric constraints is necessary in order to formulate the required boundary and/or initial conditions.

The resulting mathematical model may be linear or non-linear, depending upon the behaviour of the elements or components of the dynamic system.

3. Mathematical solution of the governing equations

The mathematical modelling of a physical vibrating system results in the formulation of the governing equations of motion. Mathematical modelling of typical systems leads to a system of differential equations of motion. The governing equations of motion of a system are solved to find the response of the system. There are many techniques available for finding the solution, namely, the standard methods for the solution of ordinary differential equations, Laplace transformation methods, matrix methods, and numerical methods. In general, exact analytical solutions are available for many linear dynamic systems, but for only a few non-linear systems. Of course, exact analytical solutions are always preferable to numerical or approximate solutions.

4. Physical interpretation of the results

The solution of the governing equations of motion for the physical system generally gives the performance. To verify the validity of the model, the predicted performance is compared with the experimental results. The model may have to be refined or a new model is developed and a new prediction compared with the experimental results. Physical interpretation of the results is an important and final step in the analysis procedure. In some situations, this may involve (a) drawing general inferences from the mathematical solution, (b) development of design curves, (c) arrive at a simple arithmetic to arrive at a conclusion (for a typical or specific problem), and (d) recommendations regarding the significance of the results and any changes (if any) required or desirable in the system involved.

1.5.1 COMPONENTS OF VIBRATING SYSTEMS

(a) Stiffness elements

Some times it requires finding out the equivalent spring stiffness values when a continuous system is attached to a discrete system or when there are a number of spring elements in the system. Stiffness of continuous elastic elements such as rods, beams, and shafts, which produce restoring elastic forces, is obtained from deflection considerations.

The stiffness coefficient of the rod (Fig. 1.3) is given by $k = \dfrac{EA}{l}$

The cantilever beam (Fig.1.4) stiffness is $k = \dfrac{3EI}{l^3}$

The torsional stiffness of the shaft (Fig.1.5) is $K = \dfrac{GJ}{l}$

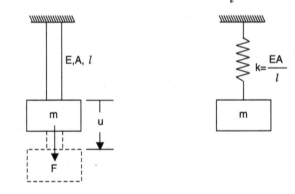

Fig.1.3 Longitudinal vibration of rods.

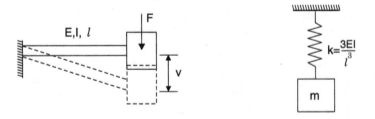

Fig.1.4 Transverse vibration of cantilever beams.

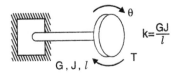

Fig. 1.5 Torsional system.

When there are several springs arranged in parallel as shown in Fig. 1.6, the equivalent spring constant is given by algebraic sum of the stiffness of individual springs. Mathematically,

$$k_{eq} = \sum_{i=1}^{n} k_i$$

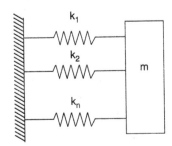

Fig. 1.6 Springs in parallel.

When the springs are arranged in series as shown in Fig. 1.7, the same force is developed in each spring and is equal to the force acting on the mass.

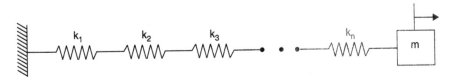

Fig. 1.7 Springs in series.

The equivalent stiffness k_{eq} is given by:

$$1/k_{eq} = \frac{1}{\displaystyle\sum_{i=1}^{n} \frac{1}{k_i}}$$

Hence, when elastic elements are in series, the reciprocal of the equivalent elastic constant is equal to the reciprocals of the elastic constants of the elements in the original system.

(*b*) **Mass or inertia elements**

The mass or inertia element is assumed to be a rigid body. Once the mathematical model of the physical vibrating system is developed, the mass or inertia elements of the system can be easily identified.

(c) Damping elements

In real mechanical systems, there is always energy dissipation in one form or another. The process of energy dissipation is referred to in the study of vibration as *damping*. A damper is considered to have neither mass nor elasticity. The three main forms of damping are *viscous damping*, *Coulomb* or *dry-friction damping*, and *hysteresis damping*. The most common type of energy-dissipating element used in vibrations study is the *viscous damper*, which is also referred to as a *dashpot*. In viscous damping, the damping force is proportional to the velocity of the body. Coulomb or dry-friction damping occurs when sliding contact that exists between surfaces in contact are dry or have insufficient lubrication. In this case, the damping force is constant in magnitude but opposite in direction to that of the motion. In dry-friction damping energy is dissipated as heat.

Solid materials are not perfectly elastic and when they are deformed, energy is absorbed and dissipated by the material. The effect is due to the internal friction due to the relative motion between the internal planes of the material during the deformation process. Such materials are known as visco-elastic solids and the type of damping which they exhibit is called as *structural* or *hysteretic damping*, or *material* or *solid damping*.

In many practical applications, several dashpots are used in combination. It is quite possible to replace these combinations of dashpots by a single dashpot of an equivalent damping coefficient so that the behaviour of the system with the equivalent dashpot is considered identical to the behaviour of the actual system.

1.6 FREE VIBRATION OF SINGLE DEGREE OF FREEDOM SYSTEMS

The most basic mechanical system is the *single-degree-of-freedom system*, which is characterized by the fact that its motion is described by a single variable or coordinates. Such a model is often used as an approximation for a generally more complex system. Excitations can be broadly divided into two types, initial excitations and externally applied forces. The behavior of a system characterized by the motion caused by these excitations is called as the *system response*. The motion is generally described by displacements.

1.6.1 FREE VIBRATION OF AN UNDAMPED TRANSLATIONAL SYSTEM

The simplest model of a vibrating mechanical system consists of a single mass element which is connected to a rigid support through a linearly elastic massless spring as shown in Fig. 1.8. The mass is constrained to move only in the vertical direction. The motion of the system is described by a single coordinate $x(t)$ and hence it has one degree of freedom (DOF).

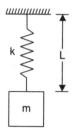

Fig. 1.8 Spring mass system.

The equation of motion for the free vibration of an undamped single degree of freedom system can be rewritten as

$$m\ddot{x}(t) + kx\,(t) = 0$$

Dividing through by m, the equation can be written in the form

$$\ddot{x}(t) + \omega_n^2 x\,(t) = 0$$

in which $\omega_n = \sqrt{k/m}$ is a real constant. The solution of this equation is obtained from the initial conditions

$$x(0) = x_0, \ \dot{x}(0) = v_0$$

where x_0 and v_0 are the initial displacement and initial velocity, respectively.

The general solution can be written as

$$x(t) = A_1 e^{i\omega_n t} + A_2 e^{-i\omega_n t}$$

in which A_1 and A_2 are constants of integration, both complex quantities. It can be finally simplified as:

$$x(t) = \frac{X}{2}\left[e^{i(\omega_n t - \phi)} + e^{-i(\omega_n t - \phi)}\right] = X \cos(\omega_n t - \phi)$$

so that now the constants of integration are X and ϕ.

This equation represents harmonic oscillation, for which reason such a system is called a *harmonic oscillator*.

There are three quantities defining the response, the *amplitude X*, the *phase angle* ϕ and the *frequency* ω_n, the first two depending on external factors, namely, the initial excitations, and the third depending on internal factors, namely, the system parameters. On the other hand, for a given system, the frequency of the response is a characteristic of the system that stays always the same, independently of the initial excitations. For this reason, ω_n is called the *natural frequency* of the harmonic oscillator.

The constants X and ϕ are obtained from the initial conditions of the system as follows:

$$X = \sqrt{x_0^2 + \left(\frac{v_0}{\omega_n}\right)^2}$$

and

$$\phi = \tan^{-1}\left[\frac{v_0}{x_0 \omega_n}\right]$$

The *time period* τ, is defined as the time necessary for the system to complete one vibration cycle, or as the time between two consecutive peaks. It is related to the natural frequency by

$$\tau = \frac{2\pi}{\omega_n} = 2\pi\sqrt{\frac{m}{k}}$$

Note that the natural frequency can also be defined as the reciprocal of the period, or

$$f_n = \frac{1}{\tau} = \frac{1}{2\pi}\sqrt{\frac{k}{m}}$$

in which case it has units of cycles per second (cps), where one cycle per second is known as one Hertz (Hz).

1.6.2 FREE VIBRATION OF AN UNDAMPED TORSIONAL SYSTEM

A mass attached to the end of the shaft is a simple torsional system (Fig. 1.9). The mass of the shaft is considered to be small in comparison to the mass of the disk and is therefore neglected.

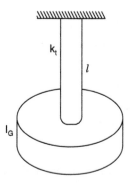

Fig. 1.9 Torsional system.

The torque that produces the twist M_t is given by

$$M_t = \frac{GJ}{l}$$

where J = the polar mass moment of inertia of the shaft $\left(J = \dfrac{\pi d^4}{32} \right.$ for a circular shaft of

diameter d $\Bigg)$

     G = shear modulus of the material of the shaft.

     l = length of the shaft.

The torsional spring constant k_t is defined as

$$k_t = \frac{T}{\theta} = \frac{GJ}{l}$$

The equation of motion of the system can be written as:

$$I_G \ddot{\theta} + k_t \theta = 0$$

The natural circular frequency of such a torsional system is $\omega_n = \left(\dfrac{k_t}{I_G} \right)^{1/2}$

The general solution of equation of motion is given by

$$\theta(t) = \theta_0 \cos \omega_n t + \frac{\dot{\theta}_0}{\omega_n} \sin \omega_n t$$

1.6.3 ENERGY METHOD

Free vibration of systems involves the cyclic interchange of kinetic and potential energy. In undamped free vibrating systems, no energy is dissipated or removed from the system. The kinetic energy T is stored in the mass by virtue of its velocity and the potential energy U is stored in the form of strain energy in elastic deformation. Since the total energy in the system is constant, the principle of conservation of mechanical energy applies. Since the mechanical energy is conserved, the sum of the kinetic energy and potential energy is constant and its rate of change is zero. This principle can be expressed as

$$T + U = \text{constant}$$

or

$$\frac{d}{dt}(T + U) = 0$$

where T and U denote the kinetic and potential energy, respectively. The principle of conservation of energy can be restated by

$$T_1 + U_1 = T_2 + U_2$$

where the subscripts 1 and 2 denote two different instances of time when the mass is passing through its static equilibrium position and select $U_1 = 0$ as reference for the potential energy. Subscript 2 indicates the time corresponding to the maximum displacement of the mass at this position, we have then

$$T_2 = 0$$

and

$$T_1 + 0 = 0 + U_2$$

If the system is undergoing harmonic motion, then T_1 and U_2 denote the maximum values of T and U, respectively and therefore last equation becomes

$$T_{max} = U_{max}$$

It is quite useful in calculating the natural frequency directly.

1.6.4 STABILITY OF UNDAMPED LINEAR SYSTEMS

The mass/inertia and stiffness parameters have an affect on the stability of an undamped single degree of freedom vibratory system. The mass and stiffness coefficients enter into the characteristic equation which defines the response of the system. Hence, any changes in these coefficient will lead to changes in the system behavior or response. In this section, the effects of the system inertia and stiffness parameters on the stability of the motion of an undamped single degree of freedom system are examined. It can be shown that by a proper selection of the inertia and stiffness coefficients, the instability of the motion of the system can be avoided. A stable system is one which executes bounded oscillations about the equilibrium position.

1.6.5 FREE VIBRATION WITH VISCOUS DAMPING

Viscous damping force is proportional to the velocity $\dot{x}$ of the mass and acting in the direction opposite to the velocity of the mass and can be expressed as

$$F = c\dot{x}$$

where c is the damping constant or coefficient of viscous damping. The differential equation of motion for free vibration of a damped spring-mass system (Fig. 1.10) is written as:

$$\ddot{x} + \frac{c}{m}\dot{x} + \frac{k}{m}x = 0$$

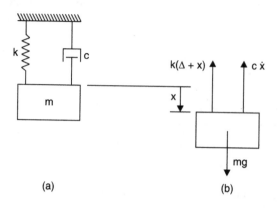

Fig. 1.10 Damped spring-mass system.

By assuming $x(t) = Ce^{st}$ as the solution, the auxiliary equation obtained is

$$s^2 + \frac{c}{m}s + \frac{k}{m} = 0$$

which has the roots

$$s_{1,\,2} = -\frac{c}{2m} \pm \sqrt{\left(\frac{c}{2m}\right)^2 - \frac{k}{m}}$$

The solution takes one of three forms, depending on whether the quantity $(c/2m)^2 - k/m$ is zero, positive, or negative. If this quantity is zero,

$$c = 2m\omega_n$$

This results in repeated roots $s_1 = s_2 = -c/2m$, and the solution is

$$x(t) = (A + Bt)e^{-(c/2m)t}$$

As the case in which repeated roots occur has special significance, we shall refer to the corresponding value of the damping constant as the *critical damping constant*, denoted by $C_c = 2m\omega_n$. The roots can be written as:

$$s_{1,\,2} = \left(-\zeta \pm \sqrt{\zeta^2 - 1}\right)\omega_n$$

where $\omega_n = (k/m)^{1/2}$ is the circular frequency of the corresponding undamped system, and

$$\zeta = \frac{c}{C_c} = \frac{c}{2m\omega_n}$$

is known as the *damping factor*.

If $\zeta < 1$, the roots are both imaginary and the solution for the motion is

$$x(t) = Xe^{-\zeta\omega_n t} \sin(\omega_d t + \phi)$$

where $\omega_d = \sqrt{1 - \zeta^2}\,\omega_n$ is called the damped circular frequency which is always less than ω, and ϕ is the phase angle of the damped oscillations. The general form of the motion is shown in Fig. 1.11. For motion of this type, the system is said to be *underdamped*.

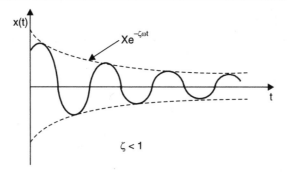

Fig. 1.11 The general form of motion.

If $\zeta = 1$, the damping constant is equal to the critical damping constant, and the system is said to be *critically damped*. The displacement is given by

$$x(t) = (A + Bt)e^{-\omega_n t}$$

The solution is the product of a linear function of time and a decaying exponential. Depending on the values of A and B, many forms of motion are possible, but each form is characterized by amplitude which decays without oscillations, such as is shown in Fig. 1.12.

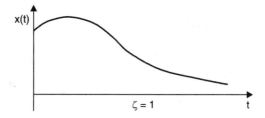

Fig. 1.12 Amplitude decaying without oscillations.

In this case $\zeta > 1$, and the system is said to be *overdamped*. The solution is given by:

$$x(t) = C_1 e^{(-\zeta + \sqrt{\zeta^2 - 1})\,\omega_n t} + C_2 e^{(-\zeta - \sqrt{\zeta^2 - 1})\,\omega_n t}$$

The motion will be non-oscillatory and will be similar to that shown in Fig. 1.13.

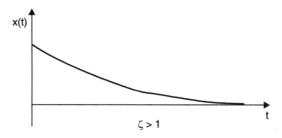

Fig. 1.13 Non-oscillatory motion.

1.6.6 LOGARITHMIC DECREMENT

The logarithmic decrement represents the rate at which the amplitude of a free damped vibration decreases. It is defined as the natural logarithm of the ratio of any two successive amplitudes.

The ratio of successive amplitudes is

$$\frac{x_i}{x_{i+1}} = \frac{Xe^{-\zeta\omega_n t_i}}{Xe^{-\zeta\omega_n (t_i + \tau_d)}} = e^{\zeta\omega_n \tau_d} = \text{constant}$$

The logarithmic decrement

$$\delta = \ln\frac{x_i}{x_{i+1}} = \ln e^{\zeta\omega_n \tau_d} = \zeta\omega_n \tau_d$$

Substituting $\tau_d = 2\pi/\omega_d = 2\pi/\omega_n \sqrt{1-\zeta^2}$ gives

$$\delta = \frac{2\pi\zeta}{\sqrt{1-\zeta^2}}$$

1.6.7 TORSIONAL SYSTEM WITH VISCOUS DAMPING

The equation of motion for such a system can be written as

$$I\ddot{\theta} + c_t\dot{\theta} + k_t\theta = 0$$

where I is the mass moment of inertia of the disc, k_t is the torsional spring constant (restoring torque for unit angular displacement), and θ is the angular displacement of the disc.

1.6.8 FREE VIBRATION WITH COULOMB DAMPING

Coulomb or dry-friction damping results when sliding contact exists between two dry surfaces. The damping force is equal to the product of the normal force and the coefficient of dry friction. The damping force is quite independent of the velocity of the motion. Consider a spring-mass system in which the mass slides on a horizontal surface having coefficient of friction f, as in Fig. 1.14.

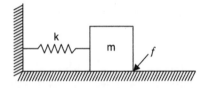

Fig. 1.14 Free vibration with coulomb damping.

The corresponding differential equations of motion of such system are

$$m\ddot{x} = -kx - F_d \quad \text{if } \ddot{x} > 0$$
$$m\ddot{x} = -kx + F_d \quad \text{if } \ddot{x} < 0$$

These differential equations and their solutions are discontinuous at the end points of their motion.

The general solution is then

$$x = A \sin \omega t + B \cos \omega t + \frac{F_d}{k} \quad (\dot{x} < 0)$$

for motion toward the left. For the initial conditions of $x = x_0$ and $\dot{x} = 0$ at $t = 0$ for the extreme position at the right, the solution becomes

$$x = \left(x_0 - \frac{F_d}{k}\right)\cos \omega t + \frac{F_d}{k} \qquad (\dot{x} < 0)$$

This holds for motion toward the left, or until $\dot{x}$ again becomes zero.

Hence the displacement is negative, or to the left of the neutral position, and has a magnitude $2F_d/k$ less than the initial displacement x_0.

A constant amplitude loss of $4F_d/k$ occurs for each cycle of motion as shown in Fig. 1.15. The motion is a linearly decaying harmonic function of time, consisting of one-half sine wave parts which are offset successively up or down by F_d/k depending on whether the motion is to the left or to the right.

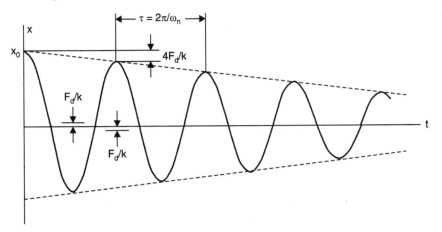

Fig. 1.15 Response of system subjected to Coulomb damping.

1.6.9 FREE VIBRATION WITH HYSTERETIC DAMPING

In general, solid materials are not perfectly elastic solid materials, in particular, metals, exhibit what is commonly referred to as *hysteretic* or *structural damping*. The hysteresis effect is due to the friction between internal planes which slip or slide as the deformations takes place. The enclosed area in the hysteresis loop is the energy loss per loading cycle. The energy loss ΔU can then be written as

$$\Delta U = \pi\beta\, kX^2$$

where β is a dimensionless *structural damping coefficient*, k is the equivalent spring constant, X is the displacement amplitude, and the factor π is included for convenience. The energy loss is a nonlinear function of the displacement.

The equivalent viscous damping constant is given by

$$c_e = \frac{\beta k}{\omega} = \beta\sqrt{mk}$$

1.7 FORCED VIBRATION OF SINGLE-DEGREE-OF-FREEDOM SYSTEMS

A mechanical or structural system is often subjected to external forces or external excitations. - The external forces may be harmonic, non-harmonic but periodic, non-periodic but having a defined form or random. The response of the system to such excitations or forces is called *forced response*. The response of a system to a harmonic excitation is called *harmonic response*. The non-periodic excitations may have a long or short duration. The response of a system to suddenly applied non-periodic excitations is called *transient response*. The sources of harmonic excitations are unbalance in rotating machines, forces generated by reciprocating machines, and the motion of the machine itself in certain cases.

1.7.1 FORCED VIBRATIONS OF DAMPED SYSTEM

Consider a viscously damped single degree of freedom spring mass system shown in Fig. 1.16, subjected to a harmonic function $F(t) = F_0 \sin \omega t$, where F_0 is the force amplitude and ω is the circular frequency of the forcing function.

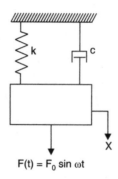

Fig. 1.16 Forced vibration of single degree of freedom system.

The equations of motion of the system is $\quad \ddot{x} + \dfrac{c}{m} \dot{x} + \dfrac{k}{m} x = \left(\dfrac{F_0}{m}\right) \sin \omega t$

The solution of the equation contains two components, complimentary function x_h and particular solution x_p. That is

$$x = x_h + x_p$$

The particular solution represents the response of the system to the forcing function. The complementary function x_h is called the *transient response* since in the presence of damping, the solution dies out. The particular integral x_p is known as the steady state solution. The steady state vibration exists long after the transient vibration disappears.

The particular solution or the steady state solution x_p can be assumed in the form

$$x_p = A_1 \sin \omega t + A_2 \cos \omega t$$

By defining $r = \dfrac{\omega}{\omega_n}, \zeta = \dfrac{c}{C_c} = \dfrac{c}{2m\omega}$, and $X_0 = F_0/k$ the amplitudes A_1 and A_2 are obtained as follows:

$$A_1 = \frac{(1 - r^2) X_0}{(1 - r^2)^2 + (2r\zeta)^2}$$

and

$$A_2 = \frac{-(2r\zeta) X_0}{(1 - r^2)^2 + (2r\zeta)^2}$$

The steady state solution x_p can be written as

$$x_p = \frac{X_0}{(1-r^2)^2 \, (2r\zeta)^2} \, [(1-r^2)\sin \omega t - (2r\zeta)\cos \omega t]$$

which can also be written as

$$x_p = \frac{X_0}{\sqrt{(1-r^2)^2 \, (2r\zeta)^2}} \sin(\omega t - \phi)$$

where X_0 is the forced amplitude and ϕ is the phase angle defined by

$$\phi = \tan^{-1}\left(\frac{2r\zeta}{1-r^2}\right)$$

It can be written in a more compact form as

$$x_p = X_0 \beta \sin(\omega t - \phi)$$

where β is known as *magnification factor*. For damped systems β is defined as

$$\beta = \frac{1}{(1-r^2)^2 + (2r\zeta)^2}$$

This forced response is called steady state solution, which is shown in Figures 1.17 and 1.18.

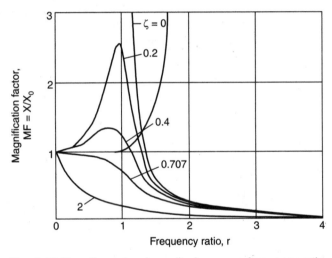

Fig. 1.17 Non-dimensional amplitude versus frequency-ratio.

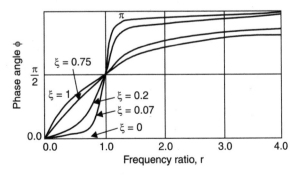

Fig. 1.18 Phase angle versus frequency-ratio.

The magnification factor β is found to be maximum when

$$r = \sqrt{1 - 2\zeta^2}$$

The maximum magnification factor is given by:

$$\beta_{max} = \frac{1}{2\zeta\sqrt{1 - \zeta^2}}$$

In the undamped systems, the particular solution reduces to

$$x_p(t) = \frac{\dfrac{F_0}{k}}{\left[1 - \left(\dfrac{\omega}{\omega_n}\right)^2\right]} \sin \omega t$$

The maximum amplitude can also be expressed as

$$\frac{X}{\delta_{st}} = \frac{1}{1 - \left(\dfrac{\omega}{\omega_n}\right)^2}$$

where $\delta_{st} = F_0/k$ denotes the static deflection of the mass under a force F_0 and is sometimes know as *static deflection* since F_0 is a constant static force. The quantity X/δ_{st} represents the ratio of the dynamic to the static amplitude of motion and is called the *magnification factor, amplification factor,* or *amplitude ratio.*

1.7.1.1 Resonance

The case $r = \dfrac{\omega}{\omega_n} = 1$, that is, when the circular frequency of the forcing function is equal to the circular frequency of the spring-mass system is referred to as *resonance*. In this case, the displacement $x(t)$ goes to infinity for any value of time t.

The amplitude of the forced response grows with time as in Fig. 1.19 and will eventually become infinite at which point the spring in the mass-spring system fails in an undesirable manner.

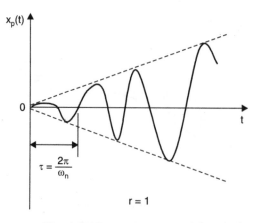

Fig. 1.19 Resonance response.

1.7.2 BEATS

The phenomenon of *beating* occurs for an undamped forced single degree of freedom spring-mass system when the forcing frequency ω is close, but not equal, to the system circular frequency ω_n. In this case, the amplitude builds up and then diminishes in a regular pattern. The phenomenon of beating can be noticed in cases of audio or sound vibration and in electric power generation when a generator is started.

1.7.3 TRANSMISSIBILITY

The forces associated with the vibrations of a machine or a structure will be transmitted to its support structure. These transmitted forces in most instances produce undesirable effects such as noise. Machines and structures are generally mounted on designed flexible supports known as *vibration isolators* or *isolators*.

In general, the amplitude of vibration reduces with the increasing values of the spring stiffness k and the damping coefficient c. In order to reduce the force transmitted to the support structure, a proper selection of the stiffness and damping coefficients must be made.

From regular spring-mass-damper model, force transmitted to the support can be written as

$$F_T = k\,x_p + c\dot{x}_p = X_0\beta\sqrt{k^2 + (c\omega)^2}\,\sin\,(\omega t - \bar{\phi})$$

where $\bar{\phi} = \phi - \phi_t$

and ϕ_t is the phase angle defined as

$$\phi_t = \tan^{-1}\left(\frac{c\omega}{k}\right) = \tan^{-1}(2r\zeta)$$

Transmitted force can also be written as:

$$F_T = F_0\beta_t\,\sin\,(\omega t - \bar{\phi})$$

where $\beta_t = \dfrac{\sqrt{1 + (2r\zeta)^2}}{\sqrt{(1 - r^2)^2 + (2r\zeta)^2}}$

The *transmissibility* β_t is defined as the ratio of the maximum transmitted force to the amplitude of the applied force. Fig. 1.20 shows a plot of β_t versus the frequency ratio r for different values of the damping factor ζ.

It can be observed from Fig. 1.20, that $\beta > 1$ for $r < \sqrt{2}$ which means that in this region the amplitude of the transmitted force is greater than the amplitude of the applied force. Also, the $r < \sqrt{2}$, the transmitted force to the support can be reduced by increasing the damping factor ζ. For $r = \sqrt{2}$, every curve passes through the point $\beta_t = 1$ and becomes asymptotic to zero as the frequency ratio is increased. Similarly, for $r > \sqrt{2}$, $\beta_t < 1$, hence, in this region the amplitude of the transmitted force is less than the amplitude of the applied force. Therefore, the amplitude of the transmitted force increases by increasing the damping factor ζ. Thus, vibration isolation is best accomplished by an isolator composed only of spring-elements for which $r > \sqrt{2}$ with no damping element used in the system.

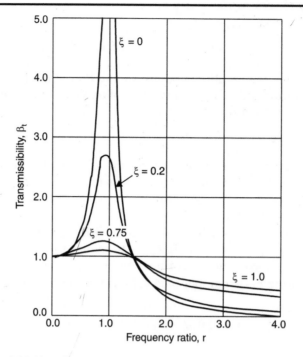

Fig. 1.20 Non-dimensional force transmitted vs. frequency ratio.

1.7.4 QUALITY FACTOR AND BANDWIDTH

The value of the amplitude ratio at resonance is also known as the *Q factor* or *Quality factor* of the system in analogy with the term used in electrical engineering applications. That is,

$$Q = \frac{1}{2\zeta}$$

The points R_1 and R_2, whereby the amplification factor falls to $Q/\sqrt{2}$, are known as *half power points*, since the power absorbed by the damper responding harmonically at a given forcing frequency is given by

$$\Delta W = \pi c \omega X^2$$

The *bandwidth* of the system is defined as the difference between the frequencies associated with the half power points R_1 and R_2 as depicted in Fig. 1.21.

It can be shown that Q-factor can be written as:

$$Q = \frac{1}{2\zeta} = \frac{\omega_n}{\omega_2 - \omega_1}$$

The quality factor Q can be used for estimating the equivalent viscous damping in a vibrating system.

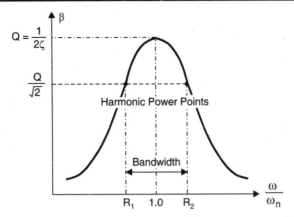

Fig. 1.21 Harmonic response curve showing half power points and bandwidth.

1.7.5 ROTATING UNBALANCE

Unbalance in many rotating mechanical systems is a common source of vibration excitation which may often lead to unbalance forces. If M is the total mass of the machine including an eccentric mass m rotating with an angular velocity ω at an eccentricity e, it can be shown that the particular solution takes the form:

$$x_p(t) = \left(\frac{me}{M}\right)\beta_r \sin(\omega t - \phi)$$

where β_r is the magnification factor which is given by

$$\beta_r = \frac{r^2}{\sqrt{(1-r^2)^2 + (2r\zeta)^2}}$$

The steady state vibration due to unbalance in rotating component is proportional to the amount of unbalance m and its distance e from the center of the rotation and increases as the square of the rotating speed. The maximum displacement of the system lags the maximum value of the forcing function by the phase angle ϕ.

1.7.6 BASE EXCITATION

In many mechanical systems such as vehicles mounted on a moving support or base, the forced vibration of the system is due to the moving support or base. The motion of the support or base causes the forces being transmitted to the mounted equipment. Fig. 1.22 shows a damped single degree of freedom mass-spring system with a moving support or base.

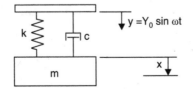

Fig. 1.22 Harmonically excited base.

The steady state solution can be written as:

$$x_p(t) = Y_0\beta_b \sin(\omega t - \phi + \phi_b),$$

where phase angle ϕ is given by $\phi = \tan^{-1}\left(\dfrac{2r\zeta}{1-r^2}\right)$ and β_b is known as the displacement

transmissibility given by: $\beta_b = \dfrac{\sqrt{1+(2r\zeta)^2}}{\sqrt{(1-r^2)^2+(2r\zeta)^2}}$

The motion of the mass relative to the support denoted by z can be written as

$$z = x - y$$

$$= \frac{Y_0 r^2}{\sqrt{(1-r^2)^2+(2r\zeta)^2}} \sin(\omega t - \phi)$$

1.7.7 RESPONSE UNDER COULOMB DAMPING

When a single-degree-of-freedom with Coulomb damping subjected to a harmonic forcing conditions, the amplitude relationship is written as:

$$X = \frac{X_0}{\sqrt{(1-r^2)^2+(4F/\pi Xk)^2}}$$

which gives $X = X_0 \dfrac{\sqrt{1-(4F/\pi F_0)^2}}{1-r^2}$

This expression for X has a real value, provided that

$$4F < \pi F_0 \quad \text{or} \quad F < \frac{\pi}{4}F_0$$

1.7.8 RESPONSE UNDER HYSTERESIS DAMPING

The steady-state motion of a single degree of freedom forced harmonically with hysteresis damping is also harmonic. The steady-state amplitude can then be determined by defining an equivalent viscous damping constant based on equating the energies.

The amplitude is given in terms of hysteresis damping coefficient β as follows

$$X = \frac{X_0}{\sqrt{(1-r^2)^2+\beta^2}}$$

1.7.9 GENERAL FORCING CONDITIONS AND RESPONSE

A general forcing function may be periodic or nonperiodic. The ground vibrations of a building structure during an earthquake, the vehicle motion when it hits a pothole, are some examples of general forcing functions. Nonperiodic excitations are referred to as *transient*. The term transient is used in the sense that nonperiodic excitations are not steady state.

1.7.10 FOURIER SERIES AND HARMONIC ANALYSIS

The Fourier series expression of a given periodic function $F(t)$ with period T can be expressed in terms of harmonic functions as

$$F(t) = \frac{a_0}{2} + \sum_{n=1}^{\infty} a_n \cos n\omega t + \sum_{n=1}^{\infty} b_n \sin n\omega t$$

where $\omega = \dfrac{2\pi}{T}$ and a_0, a_n and b_n are constants.

$F(t)$ can also be written as follows:

$$F(t) = F_0 + \sum_{n=1}^{\infty} F_n \sin(\omega_n t + \phi_n)$$

where $F_0 = a_0/2$, $F_n = \sqrt{a_n^2 + b_n^2}$, with $\omega_n = n\omega$ and $\phi_n = \tan^{-1}\left(\dfrac{a_n}{b_n}\right)$

1.8 HARMONIC FUNCTIONS

Harmonic functions are periodic functions in which all the Fourier coefficients are zeros except one coefficient.

1.8.1 EVEN FUNCTIONS

A periodic function $F(t)$ is said to be even if $F(t) = F(-t)$. A cosine function is an even function since $\cos \theta = \cos(-\theta)$. If the function $F(t)$ is an even function, then the coefficients b_m are all zeros.

1.8.2 ODD FUNCTIONS

A periodic function $F(t)$ is said to be odd if $F(t) = -F(-t)$. The sine function is an odd function since $\sin \theta = -\sin(-\theta)$. For an odd function, the Fourier coefficients a_0 and a_n are identically zero.

1.8.3 RESPONSE UNDER A PERIODIC FORCE OF IRREGULAR FORM

Usually, the values of periodic functions at discrete points in time are available in graphical form or tabulated form. In such cases, no analytical expression can be found or the direct integration of the periodic functions in a closed analytical form may not be practical. In such cases, one can find the Fourier coefficients by using a numerical integration procedure. If one divides the period of the function T into N equal intervals, then length of each such interval is $\Delta t = T/N$.

The coefficients are given by

$$a_0 = \frac{2}{N} \sum_{i=1}^{N} F(t_i)$$

$$a_n = \frac{2}{N} \sum_{i=1}^{N} F(t_i) \cos n\omega t_i$$

$$b_n = \frac{2}{N} \sum_{i=1}^{N} F(t_i) \sin n\omega t_i$$

1.8.4 RESPONSE UNDER A GENERAL PERIODIC FORCE

To find the response of a system under general periodic force, consider a single degree of freedom system shown in Fig. 1.23.

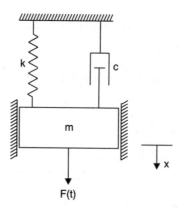

Fig. 1.23 Single degree of freedom system.

Let the periodic force $F(t)$ can be expressed in terms of harmonic functions by the use of Fourier series as follows:

$$F(t) = \frac{a_0}{2} + \sum_{n=1}^{\infty} (a_n \cos n\omega t + b_n \sin n\omega t)$$

Then steady-state solution can be written as

$$x_p(t) = \frac{a_0}{2k} + \sum_{n=1}^{\infty} \frac{a_n/k}{\sqrt{(1-r_n^2)^2 + (2\xi r_n)^2}} \cos (n\omega t - \psi_n)$$

$$+ \sum_{n=1}^{\infty} \frac{b_n/k}{\sqrt{(1-r_n^2)^2 + (2\xi r_n)^2}} \sin (n\omega t - \psi_n)$$

In most cases, the first two or three terms of this series are sufficient to describe the response of the system. If one of the harmonic frequencies $n\omega$ is close to or equal to ω, then $r \approx 1$, and the corresponding amplitude ratio can become large and resonance can occur.

1.8.5 TRANSIENT VIBRATION

When a mechanical or structural system is excited by a suddenly applied nonperiodic excitation $F(t)$, the response to such excitation is called *transient response*, as the steady-state oscillations are generally not produced.

1.8.6 UNIT IMPULSE

Impulse is time integral of the force which is finite and is written as

$$\hat{F} = \int F(t)\, dt$$

where $\hat{F}$ is the linear impulse (in pound seconds or Newton seconds) of the force.

Figure 1.24 shows an impulsive force of magnitude $F = \hat{F}/\epsilon$ acting at $t = a$ over the time interval ϵ. As ϵ approaches zero, the magnitude of the force becomes infinite but the linear impulse $\hat{F}$ is well defined.

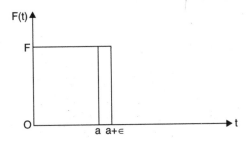

Fig. 1.24 Impulsive force.

When $\hat{F}$ is equal to unity, such a force in the limiting case ($\epsilon \to 0$) is called the *unit impulse*, or the *Direc delta function* $\delta(t-a)$, which has the following properties:

$$\delta(t-a) = 0 \quad \text{for } t \neq a$$

$$\int_0^\infty \delta(t-a)\, dt = 1$$

$$\int_0^\infty \delta(t-a)\, F(t)\, dt = F(a)$$

where $0 < a < \infty$. By using these properties, an impulsive force $F(t)$ acting at $t = a$ to produce a linear impulsive $\hat{F}$ of arbitrary magnitude can be expressed as

$$F(t) = \hat{F}\, \delta(t-a)$$

1.8.7 IMPULSIVE RESPONSE OF A SYSTEM

The response of a damped spring-mass system to an impulsive force is given by

$$x(t) = \hat{F}\, H(t)$$

where $H(t)$ is called the *impulse response function* and can be written as

$$H(t) = \frac{1}{m\omega_d}\, e^{-\zeta\omega_n t} \sin \omega_d t, \text{ where } \omega_d \text{ is damped natural frequency.}$$

If the force applied at a time $t = \tau$, this can be written as:

$$H(t-\tau) = \frac{1}{m\omega_d}\, e^{-\zeta\omega_n (t-\tau)} \sin \omega_d (t-\tau)$$

1.8.8 RESPONSE TO AN ARBITRARY INPUT

The total response is obtained by finding the integration

$$x(t) = \int_0^t F(\tau)\, H(t - \tau)\, d\tau$$

This is called the convolution integral or Duhamel's integral and is sometimes referred as the superposition integral.

1.8.9 LAPLACE TRANSFORMATION METHOD

The Laplace transformation method can be used for calculating the response of a system to a variety of force excitations, including periodic and nonperiodic. The Laplace transformation method can treat discontinuous functions with no difficulty and it automatically takes into account the initial conditions. The usefulness of the method lies in the availability of tabulated Laplace transform pairs. From the equations of motion of a single degree of freedom system subjected to a general forcing function $F(t)$, the Laplace transform of the solution $x(t)$ is given by

$$\bar{x}(s) = \frac{\overline{F}(s) + (ms + c)\, x(0) + m\dot{x}(0)}{ms^2 + cs + k}$$

The method of determining $x(t)$ given $\bar{x}(s)$ can be considered as an inverse transformation which can be expressed as

$$x(t) = L^{-1}\{\bar{x}(s)\}$$

1.9 TWO DEGREE OF FREEDOM SYSTEMS

Systems that require two independent coordinates to describe their motion are called *two degrees of freedom systems*. Some examples of two degree of freedom models of vibrating systems are shown in Fig. 1.25(*a*) and (*b*).

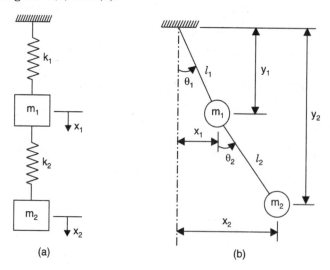

(a)   (b)

Fig. 1.25 Two degree of freedom systems.

1.9.1 EQUATIONS OF MOTION

Consider the viscously damped two-degree of freedom spring mass system shown in Fig. 1.26.

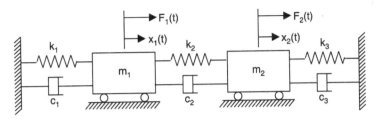

Fig. 1.26 Two-degree of freedom damped spring-mass damper.

The system is completely described by the two coordinates $x_1(t)$ and $x_2(t)$, which define the positions of the two masses m_1 and m_2, respectively, for any arbitrary time t, from the respective equilibrium positions. The external forces acting on the masses m_1 and m_2 of the system are $F_1(t)$ and $F_2(t)$ respectively.

Applying Newton's second law of motion to each of the masses m_1 and m_2 we can write the two equations of motion as:

$$m_1\ddot{x}_1(t) + (c_1 + c_2)\dot{x}_1(t) - c_2\dot{x}_2(t) + (k_1 + k_2) x_1(t) - k_2 x_2(t) = F_1(t)$$

$$m_2\ddot{x}_2(t) - c_2\dot{x}_1(t) + (c_2 + c_3)\dot{x}_2(t) - k_2 x_1(t) + (k_2 + k_3) x_2(t) = F_2(t)$$

These equations reveal that the motion of m_1 will influence the motion of mass m_2, and vice versa.

1.9.2 FREE VIBRATION ANALYSIS

Let the free vibration solution of the equations of motion be

$$x_1(t) = X_1 \cos (\omega t + \phi)$$
$$x_2(t) = X_2 \cos (\omega t + \phi)$$

where X_1 and X_2 are constants, which denote the maximum amplitudes of $x_1(t)$ and $x_2(t)$, and ϕ is the phase angle. Substituting these expressions in equations of motion leads to a characteristic determinant

$$\det \begin{bmatrix} \{-m_1\omega^2 + (k_1 + k_2)\} & -k_2 \\ -k_2 & \{m_2\omega^2 + (k_2 + k_3)\} \end{bmatrix}$$ which should be zero for consistency.

or $\quad (m_1 m_2)\omega^4 - \{(k_1 + k_2) m_2 + (k_2 + k_3) m_1\}\omega^2 + \{(k_1 + k_2)(k_2 + k_3) - k_2^2\} = 0$

This equation is known as the *frequency or characteristic equation*. The solution of this equation yields the frequencies or the characteristic values of the system.

$$\omega_1^2, \omega_2^2 = \frac{1}{2}\left\{ \frac{(k_1 + k_2)m_2 + (k_2 + k_3)m_1}{m_1 m_2} \right\}$$

$$\pm \frac{1}{2}\left[\left\{ \frac{(k_1 + k_2)m_2 + (k_2 + k_3)m_1}{m_1 m_2} \right\}^2 - 4\left\{ \frac{(k_1 + k_2)(k_2 + k_3) - k_2^2}{m_1 m_2} \right\} \right]^{1/2}$$

ω_1 and ω_2 are called the *natural frequencies* of the system.

The values of X_1 and X_2 depend on the natural frequencies ω_1 and ω_2. By denoting the values of X_1 and X_2 corresponding to ω_1 as $X_1^{(1)}$ and $X_2^{(1)}$ and those corresponding to ω_2 as $X_1^{(2)}$ and $X_2^{(2)}$:

$$r_1 = \frac{X_2^{(1)}}{X_1^{(1)}} = \frac{-m_1\omega_1^2 + (k_1 + k_2)}{k_2} = \frac{k_2}{-m_2\omega_1^2 + (k_2 + k_3)}$$

$$r_2 = \frac{X_2^{(2)}}{X_1^{(2)}} = \frac{-m_1\omega_2^2 + (k_1 + k_2)}{k_2} = \frac{k_2}{-m_2\omega_2^2 + (k_2 + k_3)}$$

The normal modes of vibration corresponding to ω_1^2 and ω_2^2 can be expressed, respectively, as

$$\{X^{(1)}\} = \begin{Bmatrix} X_1^{(1)} \\ X_2^{(1)} \end{Bmatrix} = \begin{Bmatrix} X_1^{(1)} \\ r_1 X_1^{(1)} \end{Bmatrix}$$

and

$$\{X^{(2)}\} = \begin{Bmatrix} X_1^{(2)} \\ X_2^{(2)} \end{Bmatrix} = \begin{Bmatrix} X_1^{(2)} \\ r_2 X_1^{(2)} \end{Bmatrix}$$

The vectors $\{X^{(1)}\}$ and $\{X^{(2)}\}$, which denote the normal modes of vibration, are known as the *modal vectors* of the system.

1.9.3 TORSIONAL SYSTEM

Consider the torsional system shown in Fig. 1.27, consisting of two disks on a shaft supported in frictionless bearings at the ends.

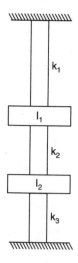

Fig. 1.27 Torsional system.

The differential equations of motion as

$$I_1\ddot{\theta}_1 + (k_1 + k_2)\theta_1 - k_2\theta_2 = 0$$

$$I_2\ddot{\theta}_2 + (k_2 + k_3)\theta_2 - k_2\theta_1 = 0$$

where k_i is the torsional stiffness of shaft i, $i = 1, 2, 3$, defined as

$$k_i = \frac{G_i J_i}{l_i}$$

where G_i is the modulus of rigidity, J_i is the polar moment of inertia, and l_i is the length of the shaft. By using the matrix notation, the differential equations of motion can be written in matrix form as

$$\begin{bmatrix} I_1 & 0 \\ 0 & I_2 \end{bmatrix} \begin{bmatrix} \ddot{\theta}_1 \\ \ddot{\theta}_2 \end{bmatrix} + \begin{bmatrix} k_1 + k_2 & -k_2 \\ -k_2 & k_2 + k_3 \end{bmatrix} \begin{bmatrix} \theta_1 \\ \theta_2 \end{bmatrix} = \begin{bmatrix} 0 \\ 0 \end{bmatrix}$$

1.9.4 COORDINATE COUPLING AND PRINCIPAL COORDINATES

The term *coupling* is used in vibration analysis to indicate a connection between equations of motion. In general an n degree of freedom vibration system requires n independent coordinates to describe completely its configuration. Often, it is quite possible to find some other set of n coordinates to describe the same configuration of the system completely. Each of these sets of n coordinates is called the *generalized coordinates*.

In the dynamic equations of motion, if the mass matrix $[M]$ is non-diagonal, then mass or *dynamic coupling* exists and if the stiffness matrix $[K]$ is non-diagonal then *stiffness* or *static coupling* exists. In general, it is possible to find a coordinate system that has neither *mass* or *dynamic coupling* nor *stiffness* or *static coupling*. Then the equations are decoupled into two independent equations and can be solved independently of the other. Such coordinates are called *principal coordinates* or *normal coordinates*.

1.9.5 FORCED VIBRATIONS

When a two degree of freedom undamped system is subjected to the harmonic forces, $F_1(t) = F_1 \sin \omega t$ and $F_2(t) = F_2 \sin \omega t$, then the amplitudes of displacement of masses is given by

$$X_1 = \frac{a_{22} F_1 - a_{12} F_2}{a_{11} a_{22} - a_{12} a_{21}}$$

and

$$X_2 = \frac{a_{11} F_2 - a_{21} F_1}{a_{11} a_{22} - a_{12} a_{21}}$$

The denominator defines the natural frequencies of the system ω_1 and ω_2. The motions of the system are coupled and hence each mass will exhibit resonance even if the resonant force acts on only one mass of the system.

For a damped two-degree of spring-mass system under external forces the solution is obtained from mechanical impedance concept.

The mechanical impedance $Z_{rs}(i\omega)$ is defined as

$$Z_{rs}(i\omega) = -\omega^2 m_{rs} + i\omega c_{rs} + k_{rs}, \quad (r, s = 1, 2)$$

1.9.6 ORTHOGONALITY PRINCIPLE

If ω_1 and ω_2 are two eigenvalues (natural frequencies) and $X^{(1)}$ and $X^{(2)}$ are the corresponding eigenvectors (natural modes) they must satisfy

$$\omega_1^2 [M] X^{(1)} = [K] X^{(1)}$$
$$\omega_2^2 [M] X^{(2)} = [K] X^{(2)}$$

Then it can be shown that

For $\omega_1 \neq \omega_2$, $[X^{(2)}]^T [M]X^{(1)} = 0$

This property is very useful, as for example to check the accuracy of computation of normal modes by its application.

1.10 MULTI-DEGREE-OF-FREEDOM SYSTEMS

A *multi-degree-of-freedom system* is defined as a system whose motion is described by more than one generalized coordinate. In general, n coordinates are needed in order to describe the motion of an n-degree-of-freedom system. Fig. 1.28 shows some examples of multi degrees of freedom systems.

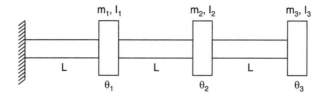

(a) Three-degree-of-freedom torsional system.

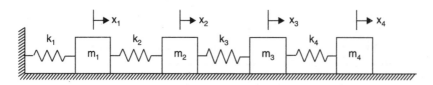

(b) Four-degree-of-freedom spring mass system.

Fig.1.28 Multi-degree of freedom systems.

An n degree-of-freedom system is governed by n coupled differential equations and has n natural frequencies. The solution of coupled differential equations can be written as the sum of a homogeneous solution and a particular solution. The free-vibration properties of the system are represented by the homogeneous solution while the particular solution represents the forced response.

1.10.1 EQUATIONS OF MOTION

Consider the motion of an n-degree of freedom system whose motion is described by the generalized coordinates, $x_1, x_2, ..., x_n$ as shown in Fig. 1.29.

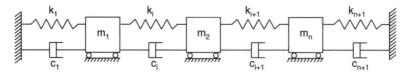

Fig. 1.29 Multi-degree of freedom system.

Applying the Newton's second law to mass m_i ($i = 1, 2, ..., n$), one can write the differential equation of motion as:

$$m_i \ddot{x}_i(t) - c_{i+1} \dot{x}_{i+1} + (c_i + c_{i+1}) \dot{x}_i - c_i \dot{x}_{i-1} - k_{i+1} x_{i+1} + (k_i + k_{i+1}) x_i - k_i x_{i-1} = 0$$

For general use, it is convenient to write this equation as in the following matrix form

$$[M]\ddot{x}(t) + [C]\dot{x}(t) + [K]x(t) = 0$$

with $[M]$, $[C]$ and $[K]$ being square matrices containing the coefficients m_{ij}, c_{ij} and k_{ij} respectively.

In this particular case, the mass-matrix is diagonal. For a different set of coordinates, $[M]$ is not necessarily diagonal.

1.10.2 STIFFNESS INFLUENCE COEFFICIENTS

For a linear system, inertial, damping and stiffness properties enter explicitly in the differential equations through the mass-coefficients m_{ij}, damping-coefficients c_{ij} and stiffness coefficients k_{ij} $(i, j = 1,2,..., n)$ respectively. Of the three, stiffness coefficients are the elastic properties causing a dynamic system to vibrate, *e.g.*, restoring-forces. *Stiffness coefficients* are also known as *stiffness influence coefficients*. Stiffness influence coefficients k_{ij} is defined as the force required at $x = x_i$ to produce a unit displacement $u_j = 1$ at point $x = x_j$ and also the displacements at all other points for which $x \neq x_j$ are zero. In other words, they define a relation between the displacement at a point and the forces acting at various other points of system. Invoking the superposition principle, the force at $x = x_i$ producing displacements u_j at $x = x_j$ $(j = 1,2,..., n)$ is

$$F_i = \sum_{j=1}^{n} k_{ij} u_j$$

1.10.3 FLEXIBILITY INFLUENCE COEFFICIENTS

Let the system be acted upon by a single-force F_j at $x = x_j$ and consider the displacement of any arbitrary point $x = x_i$ $(i = 1, 2, ..., n)$ due to force F_j. Flexibility influence coefficient is defined as the *displacement of the point $x = x_i$* due to unit force $F_j = 1$ applied at the point $x = x_j$. Invoking the principle of superposition and obtaining displacement u_i at $x = x_i$ resulting from all forces F_j $(j = 1, 2, ..., n)$ by simply summing up the individual contributions

$$U_i = \sum_{j=1}^{n} a_{ij} F_j$$

Note that the units of a_{ij} are m/N.

For a single-degree of freedom system with only one spring, the stiffness influence coefficient is merely the spring-constant, whereas the flexibility influence coefficient is its reciprocal.

1.10.4 MATRIX FORMULATION

For multi-degree of freedom systems, a more general formulation is employed. Arranging the flexibility and stiffness influence coefficients in the square matrices as

$$[a_{ij}] = [A], \text{ and } [k_{ij}] = [K]$$

where $[A]$ is the flexibility matrix and $[K]$ is the stiffness matrix.

The flexibility and stiffness matrices are the inverse of one another. Often the stiffness coefficients are easier to evaluate than the flexibility coefficients. When the stiffness matrix is singular, the flexibility matrix does not exist. This implies that the system admits rigid-body motions, in which the system undergoes no elastic deformations. This can happen when supports do not fully restrain the system from moving. Thus in the absence of adequate supports, the definition of flexibility coefficients cannot be applied, so that the coefficients are not defined.

1.10.5 INERTIA INFLUENCE COEFFICIENTS

The mass-matrix is associated with the kinetic energy. For a multi-degree of freedom system with $\dot{x}_i$ as the velocity of mass m_i ($i = 1, 2, ..., n.$), the kinetic energy is given by

$$T = \frac{1}{2} \dot{x}^T [M] \dot{x}$$

where $[M]$ is the mass-matrix or inertia matrix.

The elements of the mass-matrix m_{ij} are known as the inertia influence coefficients. The coefficients m_{ij} can be obtained using the impulse-momentum relations. The inertia influence coefficients m_{1j}, m_{2j}, ..., m_{nj} are defined as the set of impulses applied at points 1, 2, ..., n respectively, to produce a unit velocity at points 1, 2, ..., n respectively, to produce a unit velocity at point j and zero velocity at every other point. Thus, for a multi degree of freedom system, the total impulse at point i, can be found by summing up the impulses causing the velocities $\dot{x}_j$ ($j = 1, 2, ..., n$) as

$$\tilde{F} = [M]\dot{X}$$

where $[M]$ is the mass matrix, $\dot{X}$ and $\tilde{F}$ are the velocity and impulse vectors of size $n \times 1$ respectively.

1.10.6 NORMAL MODE SOLUTION

The general formulation of the differential equations governing the free-vibrations of a linear-undamped n-degree-of-freedom system can be written as

$$[M]\ddot{x} + [K]x = 0$$

where $[M]$ and $[K]$ are symmetric $n \times n$ mass and stiffness matrices respectively and x is the n-dimensional column-vector of generalized coordinates.

Free vibrations of a multi-degree-of-freedom system are initiated by the presence of an initial potential or kinetic energy.

The normal-mode solution in the form of

$$x(t) = Xe^{i\omega t}$$

where ω the frequency of vibration and X is an n-dimensional vector called a *mode shape*. Each natural frequency has at least one corresponding mode shape. The general solution is a linear superposition over all possible modes.

The frequency or eigenvalue equation is defined as

$$- \omega^2 [M]X + [K]X = 0$$

The trivial solution (X = 0) is obtained unless

$$\det[[M]^{-1} [K] - \omega^2 I] = 0$$

Thus ω^2 must be an eigenvalue of $[M]^{-1} [K]$. This form is called characteristic equation. The square of a real positive eigenvalue has two possible values, one positive and one negative. While both are used to develop the general solution, the positive square root is identified as a natural frequency. The mode shape is the corresponding eigenvector.

1.10.7 NATURAL FREQUENCIES AND MODE SHAPES

Generally in vibration problems, the characteristic equation has only real-roots since the matrices under consideration are symmetric. Assuming that all the eigenvalues of $[M]^{-1}[K]$ corresponding to the symmetric mass and stiffness matrices are non-negative. Then there exist n-real natural frequencies that can be arranged by $\omega_1 \leq \omega_2 \leq \dots \omega_n$. Each distinct eigenvalue ω_i^2, $i = 1, 2, \dots, n$, has a corresponding non-trivial eigenvector X_i, which satisfies

$$[M]^{-1}[K]X_i = \omega_i^2 X_i$$

This mode shape X_i is an n-dimensional column vector of the form

$$X_i = \begin{bmatrix} X_{i1} \\ X_{i2} \\ \vdots \\ X_{in} \end{bmatrix}$$

This mode shape is not unique. The eigenvector is unique only to arbitrary multiplicative constant. Normalization schemes exist such that the constant is chosen so the eigenvector satisfies an externally imposed condition. The algebraic complexity of the solution grows exponentially with the number of degrees of freedom. Hence, numerical methods, which do not require the evaluation of the characteristic equation, are used for systems with a large number of degrees of freedom.

1.10.8 MODE SHAPE ORTHOGONALITY

In the solution of problems involving multi-degree-of-freedom vibration, one useful fundamental relation exists between the principal modes. Consider any two principal modes of oscillation of a system of several degrees of freedom. Let these be r^{th} and s^{th} modes and the corresponding eigenvalues be ω_r^2 and ω_s^2, then it can shown that

$$\{X\}_r^T [M]\{X\}_s = 0, \quad r \neq s$$

$$\{X\}_r^T [K]\{X\}_s = 0 \quad r \neq s$$

These define the matrix form of the orthogonal relationships between principal modes of vibration. Since $[M]$ is often a diagonal matrix and $[K]$ is not, it is usually simpler to write the orthogonality matrix with respect to $[M]$. The orthogonality relation with respect to $[M]$ is written in expanded form as

$$\sum_{i=1}^{n} \sum_{j=1}^{n} m_{ij} \, x_i^r \, x_j^s = 0, \quad r \neq s$$

Thus the orthogonality relation for the principal modes of vibration is essentially a relation between the amplitudes of two principal modes. These are not necessarily successive modes but any two modes. It is convenient to normalize mode shapes by requiring that the kinetic energy scalar product of a mode shape with itself is equal to one.

1.10.9 RESPONSE OF A SYSTEM TO INITIAL CONDITIONS

Response of multi-degree-of-freedom system subjected to initial excitations $x(0)$ and $\dot{x}(0)$ in the general form can be written as

$$x(t) = \sum_{r=1}^{n} \left[U_r^T Mx(0) \cos \omega_r t + \frac{1}{\omega_r} U_r^T M\dot{x}(0) \sin \omega_r t \right] U_r$$

Here each of the natural modes can be excited independently of the other.

1.11 FREE VIBRATION OF DAMPED SYSTEMS

In the equations of free motion including viscous damping, we can assume a harmonic form for the response. Due to the presence of damping, the characteristic equation will be a polynomial that has complex conjugate roots. For a given complex conjugate eigenvalues there are conjugate eigenvectors. The normal mode method or modal analysis applies only to undamped systems or systems where the damping can be made mathematically equivalent to the mass or stiffness matrix. Sometimes damping can be ignored in the forced response of a vibrating system.

1.12 PROPORTIONAL DAMPING

For some special systems, where the damping matrix is linearly related to the mass and stiffness matrices, the simultaneous diagonalization of the stiffness and mass matrices can be accomplished along with that of the damping matrix. Such systems are called *proportional damping systems*.

Here $\qquad [C] = \alpha[K] + \beta[M]$

where α and β are constants.

Differential equations governing the free vibrations of a linear system with proportional damping can be written as

$$[M]\ddot{X} + (\alpha[K] + \beta[M])\dot{X} + [K] X = 0$$

If $\omega_1 \le \omega_2 \le \dots \omega_n$ are the natural frequencies of an undamped system whose mass-matrix is $[M]$ and stiffness matrix $[K]$ and $U_1, U_2, \dots U_n$ are the corresponding normalized mode shapes. The expansion-theorem implies that X can be written as a linear combination of the mode shape vector.

$$X = \Sigma p_i U_i$$

The matrix triple products possessing orgthogonality properties, is written as

$$[U]^T[M][U]\{\ddot{P}\} + [U]^T(\alpha[K] + \beta[M]) [U]\{\dot{P}\} + [U]^T[K][U]\{P\} = 0$$

The orthogonality of modes with respect to mass and stiffness permits the following substitutions:

$$[U]^T [M] [U] = [I]$$

and $\qquad [U]^T [K] [U] = [\text{diag } \omega^2] = [\Omega]$

The equations can now be decoupled into governing equations for each degree of freedom. Mathematically,

$$\ddot{p}_i + (\alpha\omega_i^2 + \beta) \dot{p}_i + \omega_i^2 p_i = 0$$

In this connection, modal damping ratio is defined as $\xi_i = \dfrac{1}{2}\left(\alpha\omega_i + \dfrac{\beta}{\omega_i} \right)$

The general solution for free vibration problem under $\xi_i < 1$ is given by

$$p_i(t) = A_i e^{-\xi_i \omega_i t} \sin\left(\omega_i \sqrt{1 - \xi_i^2}\ t - \phi_i\right)$$

where A_i and ϕ_i are constants determined from the initial conditions. Finally the solution is obtained in terms of generalized coordinates.

1.13 GENERAL VISCOUS DAMPING

The differential equations governing the free-vibrations of a multi-degree-of-freedom system with viscous damping are given by

$$[M]\ddot{X} + [C]\dot{X} + [K]X = 0$$

If the damping is arbitrary, then the principal coordinates of the undamped system do not uncouple the above equation. The equation can be reformulated as $2n$ first-order differential equations by writing

$$[\tilde{M}]\dot{y} + [\tilde{K}]y = 0$$

where $\quad [\tilde{M}] = \begin{bmatrix} O & [M] \\ [M] & [C] \end{bmatrix}, [\tilde{K}] = \begin{bmatrix} -[M] & O \\ O & [K] \end{bmatrix}, y = \begin{bmatrix} \dot{X} \\ X \end{bmatrix}$

If the values of γ are complex-conjugate eigenvalues of $[\tilde{M}]^{-1}[\tilde{K}]$ and ϕ is a corresponding eigenvector, then the solution takes the form as

$$y = \phi\, e^{-\gamma t}$$

1.14 HARMONIC EXCITATIONS

Differential equations governing the motion of an n-degree of freedom undamped system subject to a single-frequency excitation with all excitation terms at the same phase can be written as:

$$[M]\ddot{x} + [K]x = F \sin \omega t$$

where F is an n-dimensional vector of constant forces. A particular solution of the form is assumed as follows:

$$x(t) = U \sin \omega t$$

where U is an n-dimensional vector of undetermined coefficients.

It results in by usual method as a solution

$$U = (-\omega^2 [M] + i\omega[C] + [K])^{-1} F$$

Alternative to this method of undetermined coefficients, Laplace transform method can also be employed.

1.15 MODAL ANALYSIS FOR UNDAMPED SYSTEMS

The differential equations governing the forced vibration motion of an undamped linear n-degree-of-freedom system are

$$M\ddot{X} + KX = F$$

The method of modal analysis uses the principal coordinates of the system to uncouple this equation as follows:

$$\sum_{i=1}^{n} \ddot{p}_i \, (X_j \, MX_i) + \sum_{i=1}^{n} p_i (X_j KX_i) = X_j F$$

Application of mode shape orthogonality leads to only one non-zero term in each summation, *i.e.*, the term corresponding to $i = j$. Since the mode shapes are normalized, the following set of equations are obtained

$$\ddot{p}_j + \omega_j^2 \, p_j = g_j(t)$$

where $g_j(t) = X_j F$

If the initial conditions for p_i are both zero, then the convolution integral solution is given by

$$p_i(t) = \frac{1}{\omega_i} \int_0^t g_i(\tau) \sin \left[\omega_i (t - \tau) \right] d\tau$$

Once the solution for each p_i is obtained, the original generalized coordinates can be determined.

The same methodology can be applied to systems having proportional damping.

Here it leads to the differential equations for the principal coordinates as

$$\ddot{p}_i + 2\xi_i \omega_i \dot{p}_i + \omega_i^2 \, p_i = g_i(t)$$

where ξ_i is modal damping-ratio.

In this case, the convolution-integral solution is given by

$$p_i(t) = \frac{1}{\omega_{d_i}} \int_0^t g_i(\tau) e^{-\xi_i \omega_i (t - \tau)} \sin \omega_{d_i} (t - \tau) \, d\tau$$

where $\dfrac{1}{\omega_{d_i}} = \omega_i \sqrt{1 - \xi_i^2}$

1.16 LAGRANGE'S EQUATION

There are two general approaches to classical dynamics: *vectorial dynamics* and *analytical dynamics*. *Vectorial dynamics* is based directly on the application of Newton's second law of motion, concentrating on forces and motions. *Analytical dynamics* treats the system as a whole dealing with scalar quantities such as the kinetic and potential energies of the system. Lagrange proposed an approach, which provides a powerful and versatile method for the formulation of the equations of motion for any dynamical system. *Lagrange's equation* obtains the equation of motion in generalized coordinates approaching the system from the analytical dynamics point of view. Lagrange's equations are differential equations in which one considers the energies of the system and the work done instantaneously in time.

1.16.1 GENERALIZED COORDINATES

The coordinates used to describe the motion in each degree of freedom of a system are termed as *generalized coordinates*. They may be Cartesian, polar, cylindrical or spherical coordinates, provided any one of them can be used to describe the configuration of the system where the motion along any one coordinate direction is independent of others. But, sometimes they may not have such simple physical or geometrical meaning. For example, the deflections of a string, stretched between two points, can be expressed in the form of trigonometric Fourier series, and the coefficients of all the terms in the series can be considered as a generalized coordinate set. This is because each trigonometric function in the series may be considered as a unique degree of freedom and the coefficients describe the extent of deflection in each degree of freedom.

It is possible to transform the coordinates from any one system to the generalized coordinate system or vice versa, through coordinate transformation. Consider a mechanical system consisting of N particles whose positions are (x_i, y_i, z_i), $i = 1,2,..., N$, in a Cartesian coordinate system. The motion of the mechanical system is completely defined if the variation with time of these positions $i.e.$ $x_i = x_i(t)$, $y_i = y_i(t)$, $z_i = z_i(t)$, are known. These $3N$ coordinates completely define a representative space. If it is possible to find another set of generalized coordinates, q_i, $i = 1,2,....n$, where $n = 3N$, then these two coordinate systems are related by the following:

$$x_i(t) = x_i(q_1, q_2, ..., q_n, t)$$
$$y_i(t) = y_i(q_1, q_2, ..., q_n, t)$$
$$z_i(t) = z_i(q_1, q_2, ..., q_n, t)$$

1.17 PRINCIPLE OF VIRTUAL WORK

The principle of virtual work is essentially a statement of the static or dynamic equilibrium of a mechanical system. A *virtual displacement*, denoted by δr, is an imaginary displacement and it occurs without the passage of time. The virtual displacement being infinitesimal obeys the rules of differential calculus.

Consider a mechanical system with N particles in a three-dimensional space whose cartesian coordinates are $(x_1, y_1, z_1,...,z_n)$. Suppose the system is subject to k constraints $\phi_j(x_1, y_1, z_1,...,z_n, t) = 0, j = 1,2,...,k$. The virtual displacements $\delta x_1, \delta y_1, \delta z_1$ etc. are said to be *consistent* with the system constraints if the constraint equations are still satisfied.

The virtual work performed by the resultant force vector $\overline{F}_i$ over the virtual displacement vector δr_i of particle i is

$$\delta W = \sum_{i=1}^{N} \overline{F}_i \cdot \delta \overline{r}_i$$

When the system is in equilibrium, the resultant force acting on each particle is zero. The resultant force is the sum of the applied force and the reaction force or the constraint force. The virtual work done by all the forces in moving through an arbitrary virtual displacement consistent with the constraints is zero.

1.18 D'ALEMBERT'S PRINCIPLE

The principle of virtual work is extended to dynamics, in which form it is known as D'Alembert's principle. The principle of virtual work is extended to the dynamic case by considering the inertia forces and considering the systems to be in dynamic equilibrium.

The generalized principle of D'Alembert states that *the virtual work performed by the effective forces through infinitesimal virtual displacements compatible with the system constraints is zero.*

1.19 LAGRANGE'S EQUATIONS OF MOTION

If Q_i is called the *generalized force* in the direction of the ith generalized coordinate, T is the kinetic energy and V is potential energy, then Lagrange's equation is given by

$$\frac{d}{dt}\left(\frac{\partial T}{\partial \dot{q}_i}\right) - \frac{\partial T}{\partial q_i} + \frac{\partial V}{\partial q_i} = Q_i$$

Expressing $T - V = L$, called the *Lagrangian*, the equation can be written as

$$\frac{d}{dt}\left(\frac{\partial L}{\partial \dot{q}_i}\right) - \frac{\partial L}{\partial q_i} = Q_i$$

1.20 VARIATIONAL PRINCIPLES

An alternative approach to the study of motion is the use of variational principle, which views the motion as a whole from the beginning to the end. This involves a search for the path in the configuration space, which yields a stationary value for a certain integral. Unlike as in the case of differential equations, the initial and final points in the configuration space are fixed in this approach. The most celebrated variational principle in dynamics is the *Hamilton's principle.*

1.21 HAMILTON'S PRINCIPLE

Hamilton's principle is the most important and powerful variational principle in dynamics. It is derived from the generalized D'Alembert's principle. The generalized version of Hamilton's principle can be written as

$$\int_{t_0}^{t_1} (\delta T + \delta W)\, dt = 0 \quad \text{or} \quad \int_{t_0}^{t_1} \delta\, (T - V)\, dt = 0, \, \delta \int_{t_0}^{t_1} L\, dt = 0$$

where $L = T - V$.

The usual form of Hamilton's principle applies to a more restricted class of systems, which are called *conservative systems*. In these systems all the applied forces are derivable from a potential function $V(q, t)$.

The usual form of Hamilton's principle states that: The actual path in the configuration space followed by a holonomic system from t_0 and t_1 is such that the integral

$$I = \int_{t_0}^{t_1} L\, dt$$

is stationary with respect to any path variations, which vanish at the end points.

REFERENCES

Benaroya, H., *Mechanical Vibrations,* Prentice Hall, Upper Saddle River, NJ, 1998.

Bhat, R.B., and Dukkipati, R.V., *Advanced Dynamics,* Narosa Publishing House, New Delhi, India, 2001.

Dimarogonas, A.D., and Haddad, S.D., *Vibration for Engineers,* Prentice Hall, Englewood cliffs, NJ, 1992.

Dukkipati, R.V., *Advanced Engineering Analysis,* Narosa Publishing House, New Delhi, India, 2006.

Dukkipati, R.V., *Advanced Mechanical Vibrations,* Narosa Publishing House, New Delhi, India, 2006.

Dukkipati, R.V., and Amyot, J.R., *Computer Aided Simulation in Railway Vehicle Dynamics,* Marcel-Dekker, New York, NY, 1988.

Dukkipati, R.V., and Srinivas, J., *A Text Book of Mechanical Vibrations,* Prentice Hall of India, New Delhi, India, 2005.

Dukkipati, R.V., and Srinivas, J., *Vibrations: Problem Solving Companion,* Narosa Publishing House, New Delhi, India, 2006.

Dukkipati, R.V., *Vehicle Dynamics,* Narosa Publishing House, New Delhi, India, 2000.

Dukkipati, R.V., *Vibration Analysis,* Narosa Publishing House, New Delhi, India, 2005.

Garg, V.K., and Dukkipati, R.V., *Dynamics of Railway Vehicle Systems,* Academic Press, New York, NY, 1984.

Kelly, S.G., *Fundamentals of Mechanical Vibration,* McGraw Hill, New York, NY, 1993.

Meirovitch, L., *Elements of Vibration Analysis,* 2nd ed., McGraw Hill, New York, NY, 1986.

Newland, D.E., *Mechanical Vibration Analysis and Computation,* Longman, 1989.

Ramamurti, V., *Mechanical Vibration Practice With Basic Theory,* CRC Press, Boca Raton, FL, 2000.

Rao, J.S., and Gupta, K., *Introductory Course on Theory and Practice of Mechanical Vibrations,* Wiley Eastern, New Delhi, India, 1984.

Rao, S.S., *Mechanical Vibrations,* 3rd ed., Addison Wesley, Reading, MA, 1995.

Seto, W.W., *Theory and Problems of Mechanical Vibrations,* Schaum series, McGraw Hill, New York, NY, 1964.

Srinivasan, P., *Mechanical Vibration Analysis,* Tata McGraw Hill, New Delhi, India, 1982.

Steidel, R.F., *An Introduction to Mechanical Vibrations,* 3rd ed., Wiley, New York, NY, 1981.

Thomson, W.T., and Dahleh, M.D., *Theory of Vibrations with Applications,* 5th ed., Prentice Hall, Englewood Cliffs, NJ, 199.

Timoshenko, S., Young, D.H., and Weaver, W., *Vibration Problems in Engineering,* 5th ed., Wiley, New York, NY 1990.

Tong, K.N., *Theory of Mechanical Vibration,* Wiley, New York, NY, 1960.

Tse, F.S., Morse, I.E., and Hinkle, R.T., *Mechanical Vibrations,* Allyn and Bacon, Boston, MA, 1963.

Vierck, R.K., *Vibration Analysis,* 2nd ed., Harper & Row, New York, NY, 1979.

GLOSSARY OF TERMS

Terminology used frequently in the field of vibration analysis is compiled here from various sources including from the document prepared by ISO, TCO-108, Mechanical Vibration and Shock, American National Standards Institute Inc., New York.

Acceleration: Acceleration is a vector quantity that specifies the time rate of change of velocity.

Amplification Factor: See magnification ratio.

Amplification: The amount of power or amplitude in an electric signal. Devices such as transistors are used to create or increase amplification.

Amplitude Ratio: *Amplitude ratio* or *magnifications factor* is the ratio of the maximum force developed in the spring of a mass-spring-dashpot system to the maximum value of the exciting force.

Amplitude: Amplitude is the maximum distance from either side of the natural position that the object can travel once the object is released.

Analytical Dynamics: See analytical mechanics.

Analytical Mechanics: Analytical mechanics or variational approach to mechanics or analytical dynamics, considers the system as a whole, rather that the individual components separately, a process that excludes the reaction and constraint forces automatically.

Angular Frequency (Circular Frequency): The angular frequency of a periodic quantity, in radians per unit time, is the frequency multiplied by 2π.

Angular Impulse: The angular impulse of a constant torque T acting for a time t is the product Tt.

Angular Mechanical Impedance (Rotational Mechanical Impedance): Angular mechanical impedance is the impedance involving the ratio of torque to angular velocity. (See Impedance.)

Angular Momentum: The angular momentum of a body about its axis of rotation is the moment of its linear momentum about the axis.

Angular Motion: See rotational particle motion.

Anti-resonance: For a system in forced oscillation, anti-resonance exists at a point when any change, however small, in the frequency of excitation causes an increase in the response at this point.

Auxiliary Mass Damper (Damped Vibration Absorber): An auxiliary mass damper is a system consisting of a mass, spring, and damper which tends to reduce vibration by the dissipation of energy in the damper as a result of relative motion between the mass and the structure to which the damper is attached.

Balancing: Balancing is a procedure for adjusting the mass distribution of a rotor so that vibration of the journals, or the forces on the bearings at once-per-revolution, are reduced or controlled.

Basic Law of Nature: A basic law of nature is a physical law that applies to all physical systems regardless of the motional from which the system is constructed.

Beats: Beats are periodic variations that result from the superposition of two simple harmonic quantities of different frequencies f_1 and f_2. They involve the periodic increase and decrease of amplitude at the beat frequency.

Centrifugal Force: If a body rotates at the end of an arm, the force is provided by the tension in the arm. The reaction to this force acts at the centre of rotation and is called the centrifugal force. It represents the inertia force of the body, resisting the change in the direction of its motion.

Centripetal Acceleration: The acceleration that is directed towards the centre of rotation is called the centripetal acceleration.

Centripetal Force: A centre seeking force that causes an object to move towards the centre.

Circular Frequency: See Angular Frequency.

Circular Motion: See rotational particle motion.

Complex Function: A complex function is a function having real and imaginary parts.

Complex Vibration: Complex vibration is vibration whose components are sinusoids not harmonically related to one another. See Harmonic.

Compliance: Compliance is the reciprocal of stiffness.

Conservation of Angular Momentum: The total angular momentum of a system of masses about any one axis remains constant unless acted upon by an external torque about that axis.

Conservation of Energy: Energy can neither be created nor destroyed.

Conservation of Linear Momentum: The total momentum of a system of masses in any one direction remains constant unless acted upon by an external force in that direction.

Conservative System: In a conservative system, there is no mechanism for dissipating or adding energy.

Continuous Assumption: The continuous assumption implies that a system can be treated as a continuous piece of matter.

Continuous System: A system with an infinite number of degrees of freedom is called a continuous system or distributed parameter system.

Coulomb Damping: Coulomb damping is the damping that occurs due to dry friction when two surfaces slide against one another.

Coupled Modes: Coupled modes are modes of vibration that are not independent but which influence one another because of energy transfer from one mode to the other. (See mode of vibration.)

Coupling Factor, Electromechanical: The electromechanical coupling factor is a factor used to characterize the extent to which the electrical characteristics of a transducer are modified by a coupled mechanical system, and vice versa.

Coupling: Coupling is the term used in mechanical vibration to indicate a connection between equations of motion.

Critical Damping: Critical damping is the minimum viscous damping that will allow a displaced system to return to its initial position without oscillation.

Critical Speed: Critical speed is a speed of a rotating system that corresponds to a resonant frequency of the system.

Critical Velocity: The minimum velocity at the highest point of the loop in order to complete a cycle is called the critical velocity.

Critically Damped System: The system is said to be critically damped if the amount of damping is such that the resulting motion is on the border between the two cases of underdamped and over damped systems.

Cycle: A cycle is the complete sequence of values of a periodic quantity that occur during a period.

D'Alembert's Principle: The virtual performed by the effective forces through infinitesimal virtual displacements compatible with the system constraints is zero.

Damped Natural Frequency: The damped natural frequency is the frequency of free vibration of a damped linear system. The free vibration of a damped system may be considered periodic in the limited sense that the time interval between zero crossings in the same direction is constant, even though successive amplitudes decrease progressively. The frequency of the vibration is the reciprocal of this time interval.

Damping Ratio: The damping ratio is defined as the ratio of the actual value of the damping to the critical damping coefficient.

Damping: The process of energy dissipation is generally referred to in the study of vibrations as damping.

Degrees-of-freedom: The number of degrees-of-freedom of a mechanical system is equal to the minimum number of independent coordinates required to define completely the positions of all parts of the system at any instant of time. In general, it is equal to the number of independent displacements that are possible.

Dependent Variables: Dependent variables are the variables that describe the physical behaviour of the system.

Deterministic Excitation: If the excitation force is known at all instants of time, the excitation is said to be deterministic.

Discrete System: A system with a finite number of degrees of freedom is a discrete system.

Displacement: Displacement (or linear displacement) is the net change in a particle's position as determined from the position function.

Displacement: Displacement is a vector quantity that specifies the change of position of a body or particle and is usually measured from the mean position or position of rest. In general, it can be represented as a rotation vector or a translation vector, or both.

Distributed Parameter System: See continuous system.

Distributed Systems: Systems where mass and elasticity are considered to be distributed parameters are called distributed systems.

Driving Point Impedance: Driving point impedance is the impedance involving the ratio of force to velocity when both the force and velocity are measured at the same point and in the same direction. (See Impedance.)

Dry Friction Damping: (See Coulomb Damping.)

Duration of Shock Pulse: The duration of a shock pulse is the time required for the acceleration of the pulse to rise from some stated fraction of the maximum amplitude and to decay to this value. (See Shock Pulse.)

Dynamic Vibration Absorber (Tuned Damper): A dynamic vibration absorber is an auxiliary mass-spring system, which tends to neutralize vibration of a structure to which it is attached. The basic principle of operation is vibration out-of-phase with the vibration of such structure, thereby applying a counteracting force.

Dynamically Coupled: If an equation contains cross products of velocity, or if the kinetic energy contains cross products of velocity, that equation of motion is dynamically coupled.

Dynamics: Dynamics is the study of moving objects.

Elasticity: A material property that causes it to return to its natural state after being compressed.

Energy: Energy is the capacity to do work, mechanical energy being equal to the work done on a body in altering either its position or its velocity.

Environment: (See Natural Environments and Induced Environment.)

Equivalent System: An equivalent system is one that may be substituted for another system for the purpose of analysis. Many types of equivalence are common in vibration and shock technology: (1) equivalent stiffness; (2) equivalent damping; (3) torsional system equivalent to a translational system; (4) electrical or acoustical system equivalent to a mechanical system; etc.

Equivalent Viscous Damping: Equivalent viscous damping is a value of viscous damping assumed for the purpose of analysis of a vibratory motion, such that the dissipation of energy per cycle at resonance is the same for either the assumed or actual damping force.

Excitation (Stimulus): Excitation is an external force (or other input) applied to a system that causes the system to respond in some way.

External forces: Actions of other bodies on a rigid body are known as external forces.

Flexibility Matrix: The flexibility matrix is the inverse of the stiffness matrix.

Force: Force is a push or a pull that one body exerts on another, includes gravitational, electrostatic, magnetic and contact influences.

Forced Vibrations: Vibrations, which occurs in the presence of an external excitation, are called forced vibrations.

Foundation (Support): A foundation is a structure that supports the gravity load of a mechanical system. It may be fixed in space, or it may undergo a motion that provides excitation for the supported system.

Fraction of Critical Damping: The fraction of critical damping (damping ratio) for a system with viscous damping is the ratio of actual damping coefficient c to the critical damping coefficient cc.

Free Body Diagram Method: One method of deriving the differentiated equations of motion, referred to as the free body diagram method, involves applying conservation laws to free body diagrams of the system drawn at an arbitrary instant.

Free Vibrations: A system is undergoing free vibrations when the vibrations occur in the absence of an external excitation.

Frequency, Angular: (See Angular Frequency.)

Frequency: The frequency is the number of cycles the system executes in a period of time and is the reciprocal of the period.

Friction: Friction is a force that always resists motion or impending motion.

Fundamental Frequency: The natural frequencies can be arranged in order of increasing magnitude and the lowest frequency is referred to as the fundamental frequency.

Fundamental Mode of Vibration: The fundamental mode of vibration of a system is the mode having the lowest natural frequency.

Generalized Coordinates: A set of independent coordinates which properly and completely defines the configuration of a system and whose number is equal to the number of degrees of freedom is called the generalized coordinates.

Generalized Forces: The generalized forces are not usually actual or observable forces acting on the system, but some component of a combination of such forces.

Gravity Acceleration: Gravity is measured in terms of the acceleration a planet gives to an object on Earth. The value of gravity acceleration is 9.8 m/sec^2.

Grid Points: See Mesh Points.

Harmonic Excitations: If the excitation force is periodic the excitations is said to be harmonic.

Harmonic Motion: (See Simple Harmonic Motion.)

Harmonic Response: Harmonic response is the periodic response of a vibrating system exhibiting the characteristics of resonance at a frequency that is a multiple of the excitation frequency.

Harmonic: A harmonic is a sinusoidal quantity having a frequency that is an integral multiple of the frequency of a periodic quantity to which it is related.

Holonomic Coordinates: If each of the coordinates is independent of the others, the coordinates are known as holonomic coordinates.

Holonomic System: Systems having equations of constraint containing only coordinates or coordinates and time are called holonomic systems.

Homogenous Differential Equation: A differential equation in which all terms contain the unknown function or its derivative is known as a homogenous differentiated equation.

Hysteric Damping: The existence of hysterics loop leads to energy dissipation from the system during each cycle, which causes natural damping called hysterics damping.

Impact: An impact is a single collision of one mass in motion with a second mass, which may be either in motion or at rest.

Impedance: Mechanical impedance is the ratio of a force-like quantity to a velocity-like quantity when the arguments of the real (or imaginary) parts of the quantities increase linearly with time. Examples of force-like quantities are: force, sound pressure, voltage, and temperature. Examples of velocity-like quantities are: velocity, volume velocity, current, and heat flow. Impedance is the reciprocal of mobility.

Impulse: Impulse is the product of a force and the time during which the force is applied; more specifically, the impulse is Fdt where the force F is time dependent and equal to zero before time t_1 and after time t_2.

Impulsive Force: An impulsive force is defined as a force which has a large magnitude and acts during a very short time duration such that the time integral of the force is finite.

Impulsive Torque: A torque, which acts for a very short time, is referred to as an impulsive torque.

Independent Variables: Independent variables are the variables with which the dependent variable changes.

Influence Coefficient: An influence coefficient, denoted by $\propto_{12}$, is defined as the status deflection of the system at position 1 due to a unit force applied at position 2 when the unit force is the only force acting.

Internal Forces: Internal forces hold together parts of a rigid body.

Isolation: Isolation is a reduction in the capacity of a system to respond to an excitation, attained by the use of a resilient support. In steady-state forced vibration, isolation is expressed quantitatively as the complement of transmissibility.

Jerk: Jerk is a vector that specifies the time rate of change of acceleration; jerk is the third derivative of displacement with respect to time.

Kinematics: Kinematics is the study of a body's motion independent of the force on the body. It is a study of the geometry of motion without consideration of the causes of motion.

Kinematics: The branch of mechanics that studies the motion of objects without reference to the forces that causes the motion.

Kinetic Coefficient of Friction: Coulombs law states that the friction force is proportional to the normal force developed between the mass and the surface. The constant of proportions μ is called the kinetic coefficient of friction.

Kinetic Energy: The kinetic energy of a body is the energy it possesses due to its velocity. If a body of mass m attains a velocity v from rest under the influence of a force P and moves a distance s, then, work done by P is Ps and the kinetic energy of the body is [$1/2\ mv^2$].

Kinetics: Kinetics is the study of motion and the forces that cause motion.

Lagrange's Method: The technique known as Lagrange's method utilizes both the principle of virtual displacements and D'Alembert principle to derive the equations of motion of a vibrating system.

Lagrangian Function: See Lagrangian.

Lagrangian or Lagrangian Function: The Lagrangian or the Lagrangian function is defined as the difference between the kinetic energy and the potential energy of a system.

Line Spectrum: A line spectrum is a spectrum whose components occur at a number of discrete frequencies.

Linear Damping: With linear damping, the damping force is proportional to velocity.

Linear Differential Equation: A linear differential equation is one, which contains no products of the solution function and/or its derivatives.

Linear Mechanical Impedance: Linear mechanical impedance is the impedance involving the ratio of force to linear velocity. (See Impedance.)

Linear System: A linear system is one in which particles move only in straight line. Another name is rectilinear system.

Logarithmic Decrement: The rate of decay of amplitude expressed as the natural logarithm of the amplitude ratio is known as the logarithmic decrement.

Longitudinal Wave: A longitudinal wave in a medium is a wave in which the direction of displacement at each point of the medium is normal to the wave front.

Lumped Mass Systems: Systems that can be modelled as a combination of distinct mass and elastic elements, which possess many degrees of freedom, are often called lumped-mass systems.

Magnification Factor: The magnification factor (also known as the amplitude ratio and amplification factor) is defined as the ratio of the steady-state vibration amplitude to the pseudo-static deflection.

Mass: The mass of a body is determined by comparison with a standard mass, using a beam-type balance.

Matrix Iteration: Matrix iteration is a numerical procedure that allows determination of a system's natural frequencies and mode shapes successively, beginning with the smallest natural frequency.

Mean Square Value: The mean square value of a time function is found from the average of the squared values integrated over some time interval.

Mechanical Impedance: (See Impedance.)

Mechanical Shock: Mechanical shock is a nonperiodic excitation (e.g., a motion of the foundation or an applied force) of a mechanical system that is characterized by suddenness and severity, and usually causes significant relative displacements in the system.

Mechanical System: A mechanical system is an aggregate of matter comprising a defined configuration of mass, stiffness, and damping.

Mesh Points: In the finite difference method, the solution domain (over which the solution of the given differential equation is required) is replaced with a finite number of points, referred to as mesh or grid points.

Modal Analysis: The procedure of solving the system of simultaneous differential equations of motion by transforming them into a set of independent equations by means of the modal matrix is generally referred to as modal analysis.

Modal Matrix: The modal matrix consists of the modal vectors or characteristic vectors representing the natural modes of the system.

Modal Numbers: When the normal modes of a system are related by a set of ordered integers, these integers are called modal numbers.

Mode of Vibration: In a system undergoing vibration, a mode of vibration is a characteristic pattern assumed by the system in which the motion of every particle is simple harmonic with the same frequency. Two or more modes may exist concurrently in a multiple degree-of-freedom system.

Modulation: Modulation is the variation in the value of some parameter, which characterizes a periodic oscillation. Thus, amplitude modulation of a sinusoidal oscillation is a variation in the amplitude of the sinusoidal oscillation.

Momentum: The momentum of a body is the product of its mass and velocity.

Multiple Degree-of-freedom System: A multiple degree-of-freedom system is one for which two or more coordinates are required to define completely the position of the system at any instant.

N Degrees of Freedom System: When n independent coordinates are required to specify the positions of the masses of a system, the system is of n degrees of freedom.

Natural Environments: Natural environments are those conditions generated by the forces of nature and whose effects are experienced when the equipment or structure is at rest as well as when it is in operation.

Natural Frequencies: The positive square roots of the characteristic values or eigenvalues are called the natural frequencies of the system and represent the circular frequencies at which the system can oscillate.

Natural Frequency: Natural frequency is the frequency of free vibration of a system. For a multiple degree-of-freedom system, the natural frequencies are the frequencies of the normal modes of vibration.

Natural Modes: The eigenvectors are also referred to as modal vectors and represent physically the so-called natural modes.

Natural Motions: The free vibrations problem admits special independent solutions in which the system vibrates in any one of the natural modes. These solutions are referred to as natural motions.

Natural Vibration: If the oscillating motion about an equilibrium point is the result of a disturbing force that is applied once and then removed, the motion is known as natural (or free) vibration.

Negative Damping: Negative damping happens when energy is added to the system rather than the traditional dissipation. Such a system can be unstable.

Newtonian Mechanics or Vectorial Mechanics: In Newtonian mechanics, the equations of motion are expressed in terms of physical coordinates and forces, both quantities conveniently represented by vectors. Newtonian mechanics is often called or referred to as vectorial mechanics.

Node: The principle-mode vibration exhibits a point for which the displacement is zero at all times. Such a point is called a node.

Nonholonomic Systems: Systems having equations of constraint containing velocities are called nonholonomic systems.

Nonlinear Damping: Nonlinear damping is damping due to a damping force that is not proportional to velocity.

Nonlinear System: A system is nonlinear if its motion is governed by nonlinear differential equations.

Normal Mode of Vibration: A normal mode of vibration is a mode of vibration that is uncoupled from (*i.e.*, can exist independently of) other modes of vibration of a system. When vibration of the system is defined as an eigenvalue problem, the normal modes are the eigenvectors and the normal mode frequencies are the eigenvalues. The term "classical normal mode" is sometimes applied to the normal modes of a vibrating system characterized by vibration of each element of the system at the same frequency and phase. In general, classical normal modes exist only in systems having no damping or having particular types of damping.

Normal Modes: The process of adjusting the elements of the natural modes to render their amplitude is called normalization, and the resulting vectors are referred to as normal modes.

Normalization: It is often convenient to choose the magnitude of the modal vectors so as to reduce matrix $[m]$ to the identity matrix, which automatically reduces the matrix $[k]$ to the diagonal matrix of natural frequencies squared. This process is known as normalization.

Number of Degrees of Freedom: The number of degrees of freedom is equal to the number of coordinates required to completely specify the state of an object.

Orthonormal: If the modes are normalized, then they are called orthonormal.

Oscillation: Oscillation is the variation, usually with time, of the magnitude of a quantity with respect to a specified reference when the magnitude is alternately greater and smaller than the reference.

Overdamped System: If the damping is heavy, the motion is non-oscillatory, and the system is said to be overdamped.

Partial Node: A partial node is the point, line, or surface in a standing-wave system where some characteristic of the wave field has minimum amplitude differing from zero. The appropriate modifier should be used with the words "partial node" to signify the type that is intended; *e.g.*, displacement partial node, velocity partial node, pressure partial node.

Peak Value: The peak value generally refers to the maximum stress that the vibrating part is undergoing.

Peak-to-Peak Value: The peak-to-peak value of a vibrating quantity is the algebraic difference between the extremes of the quantity.

Period: The time it takes to complete a full cycle is called a period.

Periodic Quantity: A periodic quantity is an oscillating quantity whose values recur for certain increments of the independent variable.

Phase of a Periodic Quantity: The phase of a periodic quantity, for a particular value of the independent variable, is the fractional part of a period through which the independent variable has advanced, measured from an arbitrary reference.

Pickup: (See Transducer.)

Positive (Negative) Semi definite Matrix: A matrix whose elements are the coefficients of a positive (negative) semi definite quadratic form is said to be a positive (negative) semi definite matrix.

Positive Definite System: When both the mass matrix $[m]$ and the stiffness matrix $[k]$ are positive definite, the system is said to be positive definite system and the motion is that of undamped free vibrations.

Positive Semi definite System: When the mass matrix $[m]$ is positive definite and the stiffness matrix $[- k]$ is only positive semi definite, the system is referred to as a positive semi definite system and the motion is undamped free vibration.

Potential Energy: The potential energy of a body is the energy it possesses due to its position and is equal to the work done in raising it from some datum level. Thus the potential energy of a body of mass m at a height h above datum level is mgh.

Principle of Virtual Work: The work performed by the applied forces through infinitesimal virtual displacements compatible with the system constraints is zero.

Q (Quality Factor): The quantity Q is a measure of the sharpness of resonance or frequency selectivity of a resonant vibratory system having a single degree of freedom, either mechanical or electrical. In a mechanical system, this quantity is equal to one-half the reciprocal of the damping ratio. It is commonly used only with reference to a lightly damped system, and is then approximately equal to the following:

Quasi-Sinusoid: A function of the form $a = A \sin (21 \pi ft\text{-}theta)$ where either A or f, or both, is not a constant but may be expressed readily as a function of time. Ordinarily theta is considered constant.

Random Excitation: If the excitation force is unknown but average and standard deviations are known, the excitation is said to be random.

Random Sine Wave: (See Narrow-band Random Vibration.)

Random Vibration: Random vibration is vibration whose instantaneous magnitude is not specified for any given instant of time. The instantaneous magnitudes of a random vibration are specified only by-probability distribution functions giving the probable fraction of the total time that the magnitude (or some sequence of magnitudes) lies within a specified range. Random vibration contains no periodic or quasi-periodic constituents.

Ratio of Critical Damping: (See Fraction of Critical Damping.)

Rayleigh Method: Rayleigh method is a technique for obtaining an estimate of the fundamental frequency of a conservative mechanical system.

Rectilinear system: See linear system.

Resonance Frequency (Resonant Frequency): A frequency at which resonance exists.

Resonance: The condition where the amplitude increases without bound is called resonance.

Response Spectrum: See Shock Spectrum.

Response: The response of a device or system is the motion (or other output) resulting from an excitation (stimulus) under specified conditions.

Rigid Body: A rigid body does not deform when loaded and can be considered a combination of two or more particles that remain at as dices, finite distance from each other.

Root Mean Square Value: The root mean square (rms) value is the square root of the mean square value.

Rotational Particle Motion: Also known as angular motion and circular motion it is the motion of a particle around a circular path.

Self-induced (Self-excited) Vibration: The vibration of a mechanical system is self-induced if it results from conversion, within the system, of nonoscillatory excitation to oscillatory excitation.

Shock Absorber: A shock absorber is a device, which dissipates energy t, modifies the response of a mechanical system to applied shock.

Shock Impulse: See shock pulse.

Shock Isolator (Shock Mount.): A shock isolator is a resilient support that tends to isolate a system from a shock motion.

Shock Motion: Shock motion is an excitation involving motion of a foundation. (See Foundation and Mechanical Shock.)

Shock Mount: (See Shock Isolator.)

Shock Pulse: A shock pulse (shock impulse) is a disturbing force characterized by a rise and subsequent delay of acceleration in a very short period of time.

Shock Spectrum (Response Spectrum): A shock spectrum is a plot of the maximum response experienced by a single degree-of-freedom system, as a function of its own natural frequency, in response to an applied shock. The response may be expressed in terms of acceleration, velocity or displacement.

Shock: Shock is a transient phenomenon. Shock results in a sharp, nearly sudden change in velocity.

Shock-Pulse Duration: (See Duration of Shock Pulse.)

Simple Harmonic Motion: Simple harmonic motion is characterized by periodic oscillation about the equilibrium position.

Single Degree-of-freedom System: A single degree-of-freedom system is one for which only one coordinate is required to define completely the configuration of the system at any instant.

Sinusoidal Motion: (See Simple Harmonic Motion.)

Spectrum: A spectrum is a definition of the magnitude of the frequency components that constitute a quantity.

Spring Constant: See Spring Stiffness.

Spring Stiffness: A linear spring obeys a force-displacement law of $F = {}^*x$ where $*$ is called the spring stiffness or spring constant and has dimensions of force for length, and x is the displacement of the spring.

Spring: A spring is a flexible mechanical line between two particles in a mechanical system.

Static Deflection: Static deflection is the deflection of a mechanical system due to gravitational force alone.

Static Friction: The frictional force exerted on a stationery body is known as static friction, Coulomb friction, and fluid friction.

Statically Coupled: If an equation of motion contains cross products of coordinates, that equation of motion is statically coupled.

Steady-State Response: The steady-state response is the system response after the transient motion has decayed sufficiently.

Steady-State Vibration: Steady-state vibration exists in a system if the velocity of each particle is a continuing periodic quantity.

Stiffness: Stiffness is the ratio of change of force (or torque) to the corresponding change in translational (or rotational) deflection of an elastic element.

Strain Energy: The strain energy of a body is the energy stored when the body is deformed. If an elastic body of stiffness S is extended a distance x by a force P, then, work done is equal to the strain energy equal to $[1/2\ Sx^2]$.

Structural Damping: Structural damping, which results from within the structure, due to energy loss in the material or at joints.

Sub Harmonic Response: Sub harmonic response is a term sometimes used to denote a particular type of harmonic response which dominates the total response of the system. It frequently occurs when the excitation frequency is submultiples of the frequency of the fundamental resource.

Sub Harmonic: A sub harmonic is a sinusoidal quantity having a frequency that is integral submultiples of the fundamental frequency of a periodic quantity to which it is related.

Synchronous: Two harmonic oscillations are called synchronous if they have the same frequency (or angular velocity).

Torsional Pendulum: The system of a torsional spring and mass is referred to as a torsional pendulum.

Torsional Vibration: Torsional vibration refers to vibration of a rigid body about a specific reference axis. The displacement is measured in terms of an angular coordinate.

Torsional Spring: A torsional spring is a link in a mechanical system where application of a torque leads to an angular displacement between the ends of the torsional spring.

Transducer (Pickup): A transducer is a device which converts shock or vibratory motion into an optical, a mechanical, or most commonly to an electrical signal that is proportional to a parameter of the experienced motion.

Transducer: a device that converts an input energy into output energy. The output energy is usually a different type of energy than the input energy.

Transfer Impedance: Transfer impedance between two points is the impedance involving the ratio of force to velocity when force is measured at one point and velocity at the other point. The term transfer impedance also is used to denote the ratio of force to velocity measured at the same point but in different directions (See Impedance.)

Transient Vibration: Transient vibration is the temporarily sustained vibration of a mechanical system. It may consist of forced or free vibration or both.

Transmissibility: Transmissibility is the nondimensional ratio of the response amplitude of a system in steady-state forced vibration to the excitation amplitude. The ratio may be one of forces, displacement, velocities, or accelerations.

Transverse Wave: A transverse wave is a wave in which the direction of displacement at each point of the medium is parallel to the wave front.

Tuned System: A tuned system is the one for which the natural frequency of the vibration absorber is equal to the frequency that is to be eliminated (*i.e.*, the forcing frequency).

Uncoupled Mode: An uncoupled mode of vibration is a mode that can exist in a system concurrently with and independently of other modes.

Undamped Natural Frequency: The undamped natural frequency of a mechanical system is the frequency of free vibration resulting from only elastic and inertial forces of the system.

Underdamped System: For the motion of a system where the displacement is a harmonic function having amplitude which decays exponentially with time, the system is said to be underdamped, and the damping is below critical.

Uniform Mass Damping: A system is said to possess uniform mass damping if the damping which acts on each mass is proportional to the magnitude of the mass.

Uniform Motion: The term uniform motion means uniform velocity.

Unrestrained Systems: An unrestrained system has a rigid body mode corresponding to a natural frequency of zero.

Unstretched length: The length of a spring when it is not subjected to external forces is called its unstretched length.

Variance: Variance is the mean of the squares of the deviations from the mean value of a vibrating quantity.

Variational Approach to Mechanics: See Analytical Mechanics

Vector: A mathematical term for a quantity that has both magnitude and direction.

Vectorial Mechanics: See Newtonian Mechanics.

Velocity: Velocity is the rate of linear motion of a body in a particular direction. Velocity is a vector quantity.

Vibration Control: Vibration control is the use of vibration analysis to develop methods to eliminate or reduce unwanted vibrations or to use vibrations to protect against unwanted force or motion transmission.

Vibration Damper: A vibration damper is an auxiliary system composed of an inertia element and a viscous damper that is connected to a primary system as a means of vibration control.

Vibration Isolator: A vibration isolator is a resilient support that tends to isolate a system from steady-state excitation.

Vibration Machine: A vibration machine is a device for subjecting a mechanical system to control and reproducible mechanical vibration.

Vibration Meter: A vibration meter is an apparatus for the measurement of displacement velocity, or acceleration of a vibrating body.

Vibration Mount: (See Vibration Isolator.)

Vibration Pickup: (See Transducer.)

Vibration: Vibration is an oscillation where in the quantity is a parameter that defines the motion of a mechanical system. (See Oscillation.)

Vibratory Motion: See Vibration.

Viscous Damper: Viscous damper, which is also referred to as a dashpot is characterized by the resistive force exerted on a body moving in a viscous fluid, and hence the name.

Viscous Damping: Viscous damping is the dissipation of energy that occurs when a particle in a vibrating system is resisted by a force that has a magnitude proportional to the magnitude of the velocity of the particle and direction opposite to the direction of the particle.

Wave: A wave is a disturbance, which is propagated in a medium in such a manner that at any point in the medium the quantity serving as measure of disturbance is a function of the time, while at any instant the displacement at a point is a function of the position of the point. Any physical quantity that has the same relationship to some independent variable (usually time) that a propagated disturbance has, at a particular instant, with respect to space, may be called a wave.

Weight: The weight of a body is the force of attraction, which the earth exerts upon it and is determined by a suitably calibrated spring-type balance.

Work: Work is the product of the average force and the distance moved in the direction of the force by its point of application.

CHAPTER **2**

MATLAB Basics

2.1 INTRODUCTION

This Chapter is a brief introduction to **MATLAB** (an abbreviation of **MAT**rix **LAB**oratory) basics, registered trademark of computer software, version 4.0 or later developed by the Math Works Inc. The software is widely used in many of science and engineering fields. MATLAB is an interactive program for numerical computation and data visualization. MATLAB is supported on Unix, Macintosh, and Windows environments. For more information on MATLAB, contact **The MathWorks.Com**. A Windows version of MATLAB is assumed here. The syntax is very similar for the DOS version.

MATLAB integrates mathematical computing, visualization, and a powerful language to provide a flexible environment for technical computing. The open architecture makes it easy to use MATLAB and its companion products to explore data, create algorithms, and create custom tools that provide early insights and competitive advantages.

Known for its highly optimized matrix and vector calculations, MATLAB offers an intuitive language for expressing problems and their solutions both mathematically and visually. Typical uses include:

- Numeric computation and algorithm development
- Symbolic computation (with the built-in Symbolic Math functions)
- Modeling, simulation, and prototyping
- Data analysis and signal processing
- Engineering graphics and scientific visualization

In this chapter, we will introduce the MATLAB environment. We will learn how to create, edit, save, run, and debug m-files (ASCII files with series of MATLAB statements). We will see how to create arrays (matrices and vectors), and explore the built-in MATLAB linear algebra functions for matrix and vector multiplication, dot and cross products, transpose, determinants, and inverses, and for the solution of linear equations. MATLAB is based on the language C, but is generally much easier to use. We will also see how to program logic constructs and loops in MATLAB, how to use subprograms and functions, how to use comments (%) for explaining the programs and tabs for easy readability, and how to print and plot graphics both two and three dimensional. MATLAB's functions for symbolic mathematics are presented. Use of these functions to perform symbolic operations, to develop closed form expressions for solutions to algebraic equations, ordinary differential equations, and system of equations was presented. Symbolic mathematics can also be used to determine analytical expressions for the derivative and integral of an expression.

2.1.1 STARTING AND QUITTING MATLAB

To start MATLAB click on the MATLAB icon or type in MATLAB, followed by pressing the enter or return key at the system prompt. The screen will produce the MATLAB **prompt** >> (or EDU >>), which indicates that MATLAB is waiting for a command to be entered.

In order to quit MATLAB, type quit or exit after the prompt, followed by pressing the enter or return key.

2.1.2 DISPLAY WINDOWS

MATLAB has three display windows. They are

1. A *Command Window* which is used to enter commands and data to display plots and graphs.

2. A *Graphics Window* which is used to display plots and graphs

3. An *Edit Window* which is used to create and modify M-files. M-files are files that contain a program or script of MATLAB commands.

2.1.3 ENTERING COMMANDS

Every command has to be followed by a carriage return <cr> (enter key) in order that the command can be executed. MATLAB commands are case sensitive and *lower case* letters are used throughout.

To execute an *M-file* (such as Project_1.m), simply enter the name of the file without its extension (as in Project_1).

2.1.4 MATLAB EXPO

In order to see some of the MATLAB capabilities, enter the *demo* command. This will initiate the *MATLAB EXPO*. *MATLAB Expo* is a graphical demonstration environment that shows some of the different types of operations which can be conducted with MATLAB.

2.1.5 ABORT

In order to *abort* a command in MATLAB, hold down the control key and press c to generate a local abort with MATLAB.

2.1.6 THE SEMICOLON (;)

If a semicolon (;) is typed at the end of a command, the output of the command is not displayed.

2.1.7 TYPING %

When percent symbol (%) is typed in the beginning of a line, the line is designated as a comment. When the *enter* key is pressed, the line is not executed.

2.1.8 THE clc COMMAND

Typing *clc* command and pressing *enter* cleans the command window. Once the *clc* command is executed a clear window is displayed.

2.1.9 HELP

MATLAB has a host of built-in functions. For a complete list, refer to MATLAB user's guide or refer to the *on-line Help*. To obtain help on a particular topic in the list, *e.g.*, inverse, type *help inv*.

2.1.10 STATEMENTS AND VARIABLES

Statements have the form

```
>> variable = expression
```

The equals ("=") sign implies the assignment of the expression to the variable. For instance, to enter a 2 × 2 matrix with a variable name A, we write

```
>> A == [1 2 ; 3 4] <ret>
```

The statement is executed after the carriage return (or enter) key is pressed to display

```
A =
    1    2
    3    4
```

2.2   ARITHMETIC OPERATIONS

The symbols for arithmetic operations with scalars are summarized below in Table 2.1.

Table 2.1

Arithmetic operation	Symbol	Example
Addition	+	6 + 3 = 9
Subtraction	−	6 − 3 = 3
Multiplication	*	6 * 3 = 18
Right division	/	6/3 = 2
Left division	\	6\3 = 3/6 = 1/2
Exponentiation	^	6 ^ 3 (6^3 = 216)

2.3   DISPLAY FORMATS

MATLAB has several different screen output formats for displaying numbers. These formats can be found by typing the help command: help format in the Command Window. A few of these formats are shown in Table 2.2 for 2π.

Table 2.2 Display formats

Command	Description	Example
format short	Fixed-point with 4 decimal digits	>> 351/7 ans = 50.1429
format long	Fixed-point with 14 decimal digits	>> 351/7 ans = 50.14285714285715
format short e	Scientific notation with 4 decimal digits	>> 351/7 ans = 5.0143e+001

Command	Description	Example
format long e	Scientific notation with 15 decimal digits	>> 351/7 ans = 5.014285714285715e001
format short g	Best of 5 digit fixed or floating point	>> 351/7 ans = 50.143
format long g	Best of 15 digit fixed or floating point	>> 351/7 ans = 50.1428571428571
format bank	Two decimal digits	>> 351/7 ans = 50.14
format compact	Eliminates empty lines to allow more lines with information displayed on the screen	
format loose	Adds empty lines (opposite of compact)	

2.4   ELEMENTARY MATH BUILT-IN FUNCTIONS

MATLAB contains a number of functions for performing computations which require the use of logarithms, elementary math functions, and trigonometric math functions. List of these commonly used elementary MATLAB mathematical built-in functions are given in Tables 2.3 to 2.8.

Table 2.3 Common Math Functions

Function	Description
abs(x)	Computes the absolute value of **x**.
sqrt(x)	Computes the square root of **x**.
round(x)	Rounds **x** to the nearest integer.
fix(x)	Rounds (or truncates) **x** to the nearest integer toward 0.
floor(x)	Rounds **x** to the nearest integer toward $-\infty$.
ceil(x)	Rounds **x** to the nearest integer toward ∞.
sign(x)	Returns a value of -1 if **x** is less than 0, a value of 0 if **x** equals 0, and a value of 1 otherwise.
rem (x, y)	Returns the remainder of x/y. For example, **rem(25, 4)** is 1, and **rem(100, 21)** is 16. This function is also called a **modulus** function.
exp(x)	Computes e^x, where e is the base for natural logarithms, or approximately 2.718282.
log (x)	Computes ln **x**, the natural logarithm of **x** to the base e.
log 10(x)	Computes $\log_{10}$ **x**, the common logarithm of **x** to the base 10.

Table 2.4 Exponential functions

Function	Description
exp(x)	Exponential (e^x)
log(x)	Natural logarithm
log10(x)	Base 10 logarithm
sqrt(x)	Square root

Table 2.5 Trigonometric and hyperbolic functions

Function	Description
sin(x)	Computes the sine of **x**, where **x** is in radians.
cos(x)	Computes the cosine of **x**, where **x** is in radians.
tan(x)	Computes the tangent of **x**, where **x** is in radians.
asin(x)	Computes the arcsine or inverse sine of **x**, where **x** must be between -1 and 1. The function returns an angle in radians between $-\pi/2$ and $\pi/2$.
acos(x)	Computes the arccosine or inverse cosine of **x**, where **x** must be between -1 and 1. The function returns an angle in radians between 0 and π.
atan(x)	Computes the arctangent or inverse tangent of **x**. The function returns an angle in radians between $-\pi/2$ and $\pi/2$.
atan2(y,x)	Computes the arctangent or inverse tangent of the value y/x. The function returns an angle in radians that will be between $-\pi$ and π, depending on the signs of **x** and **y**.
sinh(x)	Computes the hyperbolic sine of **x**, which is equal to $\dfrac{e^{\mathbf{x}} - e^{-\mathbf{x}}}{2}$.
cosh(x)	Computes the hyperbolic cosine of **x**, which is equal to $\dfrac{e^{\mathbf{x}} + e^{-\mathbf{x}}}{2}$.
tanh(x)	Computes the hyperbolic tangent of **x**, which is equal to $\dfrac{\sinh \mathbf{x}}{\cosh \mathbf{x}}$.
asinh(x)	Computes the inverse hyperbolic sine of **x**, which is equal to $\ln\left(\mathbf{x} + \sqrt{\mathbf{x}^2 + 1}\right)$.
acosh(x)	Computes the inverse hyperbolic cosine of **x**, which is equal to $\ln\left(\mathbf{x} + \sqrt{\mathbf{x}^2 - 1}\right)$.
atanh(x)	Computes the inverse hyperbolic tangent of **x**, which is equal to $\ln\sqrt{\dfrac{1+\mathbf{x}}{1-\mathbf{x}}}$ for $\lvert\mathbf{x}\rvert \le 1$.

Table 2.6 Round-off functions

Function	Description	Example
round(x)	Round to the nearest integer	>> round(20/6) ans = 3
fix(x)	Round towards zero	>> fix(13/6) ans = 2
ceil(x)	Round towards infinity	>> ceil(13/5) ans = 3
floor(x)	Round towards minus infinity	>> floor(− 10/4) ans = −3
rem(x,y)	Returns the remainder after **x** is divided by **y**	>> rem(14,3) ans = 2
sign(x,y)	Signum function. Returns 1 if **x** > 0, − 1 if **x** < 0, and 0 if **x** = 0.	>> sign(7) ans = 1

Table 2.7 Complex number functions

Function	Description
conj(x)	Computes the complex **conjugate** of the complex number **x**. Thus, if **x** is equal to $a + i\,b$, then **conj(x)** will be equal to $a - i\,b$.
real(x)	Computes the real portion of the complex number **x**
imag(x)	Computes the imaginary portion of the complex number **x**.
abs(x)	Computes the absolute value of **magnitude** of the complex number **x**.
angle(x)	Computes the angle using the value of **atan2(imag(x), real(x))**; thus, the angle value is between $-\pi$ and π.

Table 2.8 Arithmetic operations with complex numbers

Operation	Result
$c_1 + c_2$	$(a_1 + a_2) + i(b_1 + b_2)$
$c_1 + c_2$	$(a_1 - a_2) + i(b_1 - b_2)$
$c_1 \bullet c_2$	$(a_1 a_2 - b_1 b_2) + i(a_1 b_2 - a_2 b_1)$
$\dfrac{c_1}{c_2}$	$\left(\dfrac{a_1 a_2 + b_1 b_2}{a_2^{\,2} + b_2^{\,2}} \right) + i \left(\dfrac{a_2 b_1 + b_2 a_1}{a_2^{\,2} + b_2^{\,2}} \right)$
$\lvert c_1 \rvert$	$\sqrt{a_1^{\,2} + b_1^{\,2}}$ (magnitude or absolute value of c_1)
$c_1{}^{*}$	$a_1 - ib_1$ (conjugate of c_1)
(Assume that $c_1 = a_1 + ib_1$ and $c_2 = a_2 + ib_2$.)	

2.5   VARIABLE NAMES

A variable is a name made of a letter or a combination of several letters and digits. Variable names can be up to 63 (in MATLAB 7) characters long (31 characters on MATLAB 6.0). MATLAB is case sensitive. For instance, *XX*, *Xx*, *xX*, and *xx* are the names of four different variables. It should be noted here that we should not use the names of a built-in functions for a variable. For instance, avoid using: sin, cos, exp, sqrt, ..., etc. Once a function name is used to define a variable, the function cannot be used.

2.6   PREDEFINED VARIABLES

MATLAB includes a number of predefined variables. Some of the predefined variables that are available for use in MATLAB programs are summarized in Table 2.9.

Table 2.9 Predefined variables

Predefined variable in MATLAB	Description
ans	Represents a value computed by an expression but not stored in variable name.
pi	Represents the number π.
eps	Represents the floating-point precision for the computer being used. This is the smallest difference between two numbers.
inf	Represents infinity which for instance occurs as a result of a division by zero. A warning message will be displayed or the value will be printed as ∞.
i	Defined as $\sqrt{-1}$ which is: $0 + 1.0000i$.
j	Same as i.
NaN	Stands for Not a Number. Typically occurs as a result of an expression being undefined, as in the case of division of zero by zero.
clock	Represents the current time in a six-element row vector containing year, month, day, hour, minute, and seconds.
date	Represents the current date in a character string format.

2.7 COMMANDS FOR MANAGING VARIABLES

Table 2.10 lists commands that can be used to eliminate variables or to obtain information about variables that have been created. The procedure is to enter the command in the Command Window and the *Enter* key is to be pressed.

Table 2.10 Commands for managing variables

Command	Description
clear	Removes all variables from the memory.
clear x, y,	Clears/removes only variables **x, y,** and **z** from the memory.
z	Lists the variables currently in the workspace.
who	Displays a list of the variables currently in the memory and their
whos	size together with information about their bytes and class.

2.8 GENERAL COMMANDS

In Tables 2.11 to 2.15 the useful general commands on on-line help, workspace information, directory information, and general information are given.

Table 2.11 On-line help

Function	Description
help	Lists topics on which help is available.
helpwin	Opens the interactive help window.
helpdesk	Opens the web browser based help facility.
help *topic*	Provides help on *topic*.
lookfor *string*	Lists help topics containing *string*.
demo	Runs the demo program.

Table 2.12 Workspace information

Function	Description
who	Lists variables currently in the workspace.
whos	Lists variables currently in the workspace with their size.
what	Lists m-, mat-, and mex-files on the disk.
clear	Clears the workspace, all variables are removed.
clear $x\ y\ z$	Clears only variables x, y, and z.
clear all	Clears all variables and functions from workspace.
mlock fun	Locks function fun so that **clear** cannot remove it.
munlock fun	Unlocks function fun so that **clear** can remove it.
clc	Clears command window, command history is lost.
home	Same as **clc.**
clf	Clears figure window.

Table 2.13 Directory information

Function	Description
pwd	Shows the current working directory.
cd	Changes the current working directory.
dir	Lists contents of the current directory.
ls	Lists contents of the current directory, same as **dir.**
path	Gets or sets MATLAB search path.
editpath	Modifies MATLAB search path.
copyfile	Copies a file.
mkdir	Creates a directory.

Table 2.14 General information

Function	Description
computer	Tells you the computer type your are using.
clock	Gives you wall clock time and date as a vector.
date	Tells you the date as a string.
more	Controls the paged output according to the screen size.
ver	Gives the license and the version information about MATLAB installed on your computer.
bench	Benchmarks your computer on running MATLAB compared to other computers.

Table 2.15 Termination

Function	Description
c (Control –c)	Local abort, kills the current command execution.
quit	Quits MATLAB.
exit	Same as **quit.**

2.9 ARRAYS

An array is a list of numbers arranged in rows and/or columns. A one-dimensional array is a row or a column of numbers and a two-dimensional array has a set of numbers arranged in rows and columns. An array operation is performed *element-by-element*.

2.9.1 ROW VECTOR

A vector is a row or column of elements.

In a row vector the elements are entered with a space or a comma between the elements inside the square brackets. For example,

$$x = [7 \ -1 \ 2 \ -5 \ 8]$$

2.9.2 COLUMN VECTOR

In a column vector the elements are entered with a semicolon between the elements inside the square brackets. For example,

$$x = [7; \ -1; \ 2; \ -5; \ 8]$$

2.9.3 MATRIX

A matrix is a two-dimensional array which has numbers in rows and columns. A matrix is entered row-wise with consecutive elements of a row separated by a space or a comma, and the rows separated by semicolons or carriage returns. The entire matrix is enclosed within square brackets. The elements of the matrix may be real numbers or complex numbers. For example to enter the matrix,

$$A = \begin{bmatrix} 1 & 3 & -4 \\ 0 & -2 & 8 \end{bmatrix}$$

The MATLAB input command is

$$A = [1 \ 3 \ -4 \ ; 0 \ -2 \ 8]$$

Similarly for complex number elements of a matrix B

$$B = \begin{bmatrix} -5x & \ln 2x + 7\sin 3y \\ 3i & 5 - 13i \end{bmatrix}$$

The MATLAB input command is

$$B = [-5*x \quad \log(2*x) + 7*\sin(3*y); \ 3i \quad 5 - 13i]$$

2.9.4 ADDRESSING ARRAYS

A colon can be used in MATLAB to address a range of elements in a vector or a matrix.

2.9.4.1 Colon for a vector

Va$(:)$ – refers to all the elements of the vector **Va** (either a row or a column vector).

Va$(m:n)$ – refers to elements m through n of the vector **Va**.

For instance

```
>>    V = [2   5   -1   11   8   4   7   -3   11]
>>    u = V (2:8)
u =
      5   -1   11   8   4   7   -3   11
```

2.9.4.2 Colon for a matrix

Table 2.16 gives the use of a colon in addressing arrays in a matrix.

Table 2.16 Colon use for a matrix

Command	Description
A(:, n)	Refers to the elements in all the rows of a column n of the matrix A.
A(n, :)	Refers to the elements in all the columns of row n of the matrix A.
A(:, m:n)	Refers to the elements in all the rows between columns m and n of the matrix A.
A(m:n, :)	Refers to the elements in all the columns between rows m and n of the matrix A.
A(m:n, p:q)	Refers to the elements in rows m through n and columns p through q of the matrix A.

2.9.5 ADDING ELEMENTS TO A VECTOR OR A MATRIX

A variable that exists as a vector, or a matrix, can be changed by adding elements to it. Addition of elements is done by assigning values of the additional elements, or by appending existing variables. Rows and/or columns can be added to an existing matrix by assigning values to the new rows or columns.

2.9.6 DELETING ELEMENTS

An element, or a range of elements, of an existing variable can be deleted by reassigning blanks to these elements. This is done simply by the use of square brackets with nothing typed in between them.

2.9.7 BUILT-IN FUNCTIONS

Some of the built-in functions available in MATLAB for managing and handling arrays as listed in Table 2.17.

Table 2.17 Built-in functions for handling arrays

Function	Description	Example
length(A)	Returns the number of elements in the vector A.	>> A = [5 9 2 4]; >> length(A) ans = 4

Function	Description	Example

size(A)	Returns a row vector $[m, n]$, where m and n are the size $m \times n$ of the array A.	>> A = [2 3 0 8 11 ; 6 17 5 7 1]
		A =
		   2 3 0 8 11
		   6 17 5 7 1
		>> size(A)
		ans =
		   2 5

reshape(A, m, n)	Rearrange a matrix A that has r rows and s columns to have m rows and n columns. r times s must be equal to m times n.	>> A = [3 1 4 ; 9 0 7]
		A =
		   3 1 4
		   9 0 7
		>> B = reshape(A, 3, 2)
		B =
		   3 0
		   9 4
		   1 7

diag(v)	When v is a vector, creates a square matrix with the elements of v in the diagonal	>> v = [3 2 1];
		>> A = diag(v)
		A =
		   3 0 0
		   0 2 0
		   0 0 1

diag(A)	When A is a matrix, creates a vector from the diagonal elements of A.	>> A = [1 8 3 ; 4 2 6 ; 7 8 3]
		A =
		   1 8 3
		   4 2 6
		   7 8 3
		>> vec = diag(A)
		vec =
		   1
		   2
		   3

2.10 OPERATIONS WITH ARRAYS

We consider here matrices that have more than one row and more than one column.

2.10.1 ADDITION AND SUBTRACTION OF MATRICES

The addition (the sum) or the subtraction (the difference) of the two arrays is obtained by adding or subtracting their corresponding elements. These operations are performed with arrays of identical size (same number of rows and columns).

For example if A and B are two arrays (2×3 matrices).

$$A = \begin{bmatrix} a_{11} & a_{12} & a_{13} \\ a_{21} & a_{22} & a_{23} \end{bmatrix} \text{ and } B = \begin{bmatrix} b_{11} & b_{12} & b_{13} \\ b_{21} & b_{22} & b_{23} \end{bmatrix}$$

Then, the matrix addition ($A + B$) is obtained by adding A and B is

$$\begin{bmatrix} a_{11} + b_{11} & a_{12} + b_{12} & a_{13} + b_{13} \\ a_{21} + b_{21} & a_{22} + b_{22} & a_{23} + b_{23} \end{bmatrix}$$

2.10.2 DOT PRODUCT

The dot product is a scalar computed from two vectors of the same size. The scalar is the sum of the products of the values in corresponding positions in the vectors.

For n elements in the vectors A and B:

$$\text{dot product} = A \bullet B = \sum_{i=1}^{n} a_i b_i$$

dot(A, B) computes the dot product of A and B. If A and B are matrices, the dot product is a row vector containing the dot products for the corresponding columns of A and B.

2.10.3 ARRAY MULTIPLICATION

The value in position $c_{i,j}$ of the product C of two matrices, A and B, is the dot product of row i of the first matrix and column j of the second matrix:

$$c_{i,j} = \sum_{k=1}^{n} a_{i,k} b_{k,j}$$

2.10.4 ARRAY DIVISION

The division operation can be explained by means of the identity matrix and the inverse matrix operation.

2.10.5 IDENTITY MATRIX

An identity matrix is a square matrix in which all the diagonal elements are 1's, and the remaining elements are 0's. If a matrix A is square, then it can be multiplied by the identity matrix, I, from the left or from the right:

$$AI = IA = A$$

2.10.6 INVERSE OF A MATRIX

The matrix B is the inverse of the matrix A if when the two matrices are multiplied the product is the identity matrix. Both matrices A and B must be square and the order of multiplication can be AB or BA.

$$AB = BA = I$$

2.10.7 TRANSPOSE

The transpose of a matrix is a new matrix in which the rows of the original matrix are the columns of the new matrix. The transpose of a given matrix A is denoted by A^T. In MATLAB, the transpose of the matrix A is denoted by A'.

2.10.8 DETERMINANT

A determinant is a scalar computed from the entries in a square matrix. For a 2×2 matrix A, the determinant is

$$|A| = a_{11}\, a_{22} - a_{21}\, a_{12}$$

MATLAB will compute the determinant of a matrix using the **det** function:

det(A)   Computes the determinant of a square matrix A.

2.10.9 ARRAY DIVISION

MATLAB has two types of array division, which are the left division and the right division.

2.10.10 LEFT DIVISION

The left division is used to solve the matrix equation $Ax = B$ where x and B are column vectors. Multiplying both sides of this equation by the inverse of A, A^{-1}, we have

$$A^{-1}Ax = A^{-1}B$$

or $$Ix = x = A^{-1}B$$

Hence $$x = A^{-1}B$$

In MATLAB, the above equation is written by using the left division character:

$$x = A\backslash B$$

2.10.11 RIGHT DIVISION

The right division is used to solve the matrix equation $xA = B$ where x and B are row vectors. Multiplying both sides of this equation by the inverse of A, A^{-1}, we have

$$x \bullet A\,A^{-1} = B \bullet A^{-1}$$

or $$x = B \bullet A^{-1}$$

In MATLAB, this equation is written by using the right division character:

$$x = B/A$$

2.10.12 EIGENVALUES AND EIGENVECTORS

Consider the following equation,

$$AX = \lambda X \tag{2.1}$$

where A is an $n \times n$ square matrix, X is a column vector with n rows and λ is a scalar.

The values of λ for which X are nonzero are called the *eigenvalues* of the matrix A, and the corresponding values of X are called the *eigenvectors* of the matrix A.

Eq. (2.1) can also be used to find the following equation

$$(A - \lambda I)X = 0 \tag{2.2}$$

where I is an $n \times n$ identity matrix. Eq. (2.2) corresponding to a set of homogeneous equations and has nontrivial solutions only if the determinant is equal to zero, or

$$|A - \lambda I| = 0 \tag{2.3}$$

Eq. (2.3) is known as the *characteristic equation* of the matrix A. The solution to Eq. (2.3) gives the eigenvalues of the matrix A.

MATLAB determines both the eigenvalues and eigenvectors for a matrix A.

eig(A)   Computes a column vector containing the eigenvalues of A.

[Q, d] = eig(A) Computes a square matrix Q containing the eigenvectors of A as columns and a square matrix d containing the eigenvlaues (λ) of A on the diagonal. The values of Q and d are such that $Q*Q$ is the identity matrix and $A*X$ equals λ times X.

Triangular factorization or lower-upper factorization: Triangular or lower-upper factorization expresses a square matrix as the product of two triangular matrices – a lower triangular matrix and an upper triangular matrix. The **lu** function in MATLAB computes the LU factorization:

[L, U] = lu(A) Computes a permuted lower triangular factor in L and an upper triangular factor in U such that the product of L and U is equal to A.

QR factorization: The QR factorization method factors a matrix A into the product of an orthonormal matrix and an upper-triangular matrix. The **qr** function is used to perform the QR factorization in MATLAB:

[Q, R] = qr(A) Computes the values of Q and R such that $A = QR.Q$ will be an orthonormal matrix, and R will be an upper triangular matrix.

For a matrix A of size $m \times n$, the size of Q is $m \times m$, and the size of R is $m \times n$.

Singular Value Decomposition (SVD): Singular value decomposition decomposes a matrix A (size $m \times n$) into a product of three matrix factors.

$$A = USV$$

where U and V are orthogonal matrices and S is a diagonal matrix. The size of U is $m \times m$, the size of V is $n \times n$, and the size of S is $m \times n$. The values on the diagonal matrix S are called singular values. The number of non-zero singular values is equal to the rank of the matrix.

The SVD factorization can be obtained using the **svd** function:

[U, S, V] = svd(A) Computes the factorization of A into the product of three matrices, USV, where U and V are orthogonal matrices and S is a diagonal matrix.

svd(A) Returns the diagonal elements of S, which are the singular values of A.

2.11 ELEMENT-BY-ELEMENT OPERATIONS

Element-by-element operations can only be done with arrays of the same size. Element-by-element multiplication, division, and exponentiation of two vectors or matrices is entered in MATLAB by typing a period in front of the arithmetic operator. Table 2.18 lists these operations.

Table 2.18 Element-by-element operations

Arithmetic operators	
Matrix operators	*Array operators*
+ Addition	+ Addition
– Subtraction	– Subtraction
* Multiplication	•* Array multiplication
^ Exponentiation	•^ Array exponentiation
/ Left division	•/ Array left division
\ Right division	•\ Array right division

2.11.1 BUILT-IN FUNCTIONS FOR ARRAYS

Table 2.19 lists some of the many built-in functions available in MATLAB for analyzing arrays.

Table 2.19 MATLAB built-in array functions

Function	Description	Example
mean(A)	If A is a vector, returns the mean value of the elements	>> A = [3 7 2 16]; >> mean(A) ans = 14
C = max(A)	If A is a vector, C is the largest element in A. If A is a matrix, C is a row vector containing the largest element of each column of A.	>> A = [3 7 2 16 9 5 18 13 0 4]; >> C = max(A) C = 18

[d, n] = max(A)	If A is a vector, d is the largest element in A, n is the position of the element (the first if several have the max value).	>> [d, n] = max(A)
		d =
		    18
		n =
		    7

Table 2.19 MATLAB built-in array functions (continued)

Function	Description	Example
min(A)	The same as **max(A),** but for the smallest element.	>> A = [3 7 2 16]; >> min(A) ans = 2
[d, n] = min(A)	The same as **[d, n] = max(A)**, but for the smallest element.	
sum(A)	If A is a vector, returns the sum of the elements of the vector.	>> A = [3 7 2 16]; >> sum(A) ans = 28
sort(A)	If A is a vector, arranges the elements of the vector in ascending order.	>> A = [3 7 2 16]; >> sort(A) ans = 2 3 7 16
median(A)	If A is a vector, returns the median value of the elements of the vector.	>> A = [3 7 2 16]; >> median(A) ans = 5

Function	Description	Example
std(A)	If A is a vector, returns the standard deviation of the elements of the vector.	>> A = [3 7 2 16]; >> std(A) ans = 6.3770
det(A)	Returns the determinant of a square matrix A.	>> A = [1 2 ; 3 4]; >> det(A) ans = − 2
dot(a, b)	Calculates the scalar (dot) product of two vectors a and b. The vector can each be row or column vectors.	>> a = [5 6 7]; >> b = [4 3 2]; >> dot(a, b) ans = 52
cross(a, b)	Calculates the cross product of two vectors a and b, $(a \times b)$. The two vectors must have 3 elements	>> a = [5 6 7]; >> b = [4 3 2]; >> cross(a, b) ans = − 9 18 − 9

inv(A)	Returns the inverse of a square matrix A.	>> a = [1 2 3; 4 6 8; − 1 2 3];
		>> inv(A)
		ans =
		    − 0.5000 0.0000 − 0.5000
		    − 5.0000 1.5000 1.0000
		      3.5000 − 1.0000 − 0.5000

2.12 RANDOM NUMBERS GENERATION

There are many physical processes and engineering applications that require the use of *random numbers* in the development of a solution.

MATLAB has two commands *rand* and *rand n* that can be used to assign random numbers to variables.

The *rand* command: The *rand* command generates uniformly distributed over the interval [0, 1]. A *seed value* is used to initiate a random sequence of values. The seed value is initially set to zero. However, it can be changed with the *seed* function.

The command can be used to assign these numbers to a scalar, a vector, or a matrix, as shown in Table 2.20.

Table 2.20 The rand command

Command	Description	Example
rand	Generates a single random number between 0 and 1.	>> rand ans = 0.9501
rand(1, n)	Generates an n elements row vector of random numbers between 0 and 1.	>> a = rand(1, 3) a = 0.4565 0.0185 0.8214

rand(n)	Generates an $n \times n$ matrix with random numbers between 0 and 1.	>> b = rand(3)
		b =
		  0.7382 0.9355 0.8936
		  0.1763 0.9165 0.0579
		  0.4057 0.4103 0.3529

rand(m, n)	Generates an $m \times n$ matrix with random numbers between 0 and 1.	>> c = rand(2, 3)
		c =
		  0.2028 0.6038 0.1988
		  0.1987 0.2722 0.0153

randperm(n)	Generates a row vector with n elements that are random permutation of integers 1 through n.	>> randperm(7) ans = 5 2 4 7 1 6 3

2.12.1 THE RANDOM COMMAND

MATLAB will generate Gaussian values with a mean of zero and a variance of 1.0 if a normal distribution is specified. The MATLAB functions for generating Gaussian values are as follows:

randn(n)   Generates an $n \times n$ matrix containing Gaussian (or normal) random numbers with a mean of 0 and a variance of 1.

Randn(m, n)   Generates an $m \times n$ matrix containing Gaussian (or normal) random numbers with a mean of 0 and a variance of 1.

2.13 POLYNOMIALS

A *polynomial* is a function of a single variable that can be expressed in the following form:

$$f(x) = a_0 x^n + a_1 x^{n-1} + a_2 x^{n-2} + \ldots + a_{n-1} x^1 + a_n$$

where the variable is x and the coefficients of the polynomial are represented by the values a_0, a_1, ... and so on. The *degree* of a polynomial is equal to the largest value used as an exponent.

A vector represents a polynomial in MATLAB. When entering the data in MATLAB, simply enter each coefficient of the polynomial into the vector in descending order. For example, consider the polynomial

$$5s^5 + 7s^4 + 2s^2 - 6s + 10$$

To enter this into MATLAB , we enter this as a vector as

```
>>x = [5  7  0  2  -6    10]
x =
       5   7   0   2   -6   10
```

It is necessary to enter the coefficients of all the terms.

MATLAB contains functions that perform polynomial multiplication and division, which are listed below:

conv(a, b)   Computes a coefficient vector that contains the coefficients of the product of polynomials represented by the coefficients in **a** and **b**. The vectors **a** and **b** do not have to be the same size.

[q, r] = deconv(n, d) Returns two vectors. The first vector contains the coefficients of the quotient and the second vector contains the coefficients of the remainder polynomial.

The MATLAB function for determining the roots of a polynomial is the roots function:

root(a)   Determines the roots of the polynomial represented by the coefficient vector **a**.

The roots function returns a column vector containing the roots of the polynomial; the number of roots is equal to the degree of the polynomial. When the roots of a polynomial are known, the coefficients of the polynomial are determined when all the linear terms are multiplied, we can use the **poly** function:

poly(r)   Determines the coefficients of the polynomial whose roots are contained in the vector **r**.

The output of the function is a row vector containing the polynomial coefficients.

The value of a polynomial can be computed using the *polyval* function, **polyval (a, x)**. It evaluates a polynomial with coefficients **a** for the values in **x**. The result is a matrix the same size ad **x**. For instance, to find the value of the above polynomial at $s = 2$,

```
>>x = polyval([5   7  0    2  -6  10], 2)
x =
       278
```

To find the roots of the above polynomial, we enter the command **roots (a)** which determines the roots of the polynomial represented by the coefficient vector **a**.

```
>>roots([5   7  0    2  -6  10])
ans =
       -1.8652
       -0.4641 + 1.0832i
       -0.4641 - 1.0832i
        0.6967 + 0.5355i
        0.6967 - 0.5355i
% or
>> x = [5 7 0 2 -6 10]
x =
       5     7    0    2    -6    10
>> r = roots(x)
```

```
r =
        -1.8652
        -0.4641 + 1.0832i
        -0.4641 - 1.0832i
         0.6967 + 0.5355i
         0.6967 - 0.5355i
```

To multiply two polynomials together, we enter the command *conv*.

The polynomials are: $x = 2x + 5$ and $y = x^2 + 3x + 7$

```
>>x = [2  5];
>>y = [1   3   7];
>>z = conv(x, y)
z =
        2   11   29   35
```

To divide two polynomials, we use the command *deconv*.

```
z = [2 11 29 35]; x = [2 5]
   >> [g, t] = deconv (z, x)
   g = 1  3  7
   t = 0     0     0     0
```

2.14 SYSTEM OF LINEAR EQUATIONS

A system of equations is nonsingular if the matrix **A** containing the coefficients of the equations is nonsingular. A system of nonsingular simultaneous linear equations (**AX = B**) can be solved using two methods:

(*a*) Matrix Division Method.

(*b*) Matrix Inversion Method.

2.14.1 MATRIX DIVISION

The solution to the matrix equation **AX = B** is obtained using matrix division, or **X = A/B**. The vector **X** then contains the values of **x**.

2.14.2 MATRIX INVERSE

For the solution of the matrix equation **AX = B**, we premultiply both sides of the equation by **A**$^{-1}$.

$$\mathbf{A^{-1}AX = A^{-1}B}$$

or $$\mathbf{IX = A^{-1}B}$$

where **I** is the identity matrix.

Hence $$\mathbf{X = A^{-1}B}$$

In MATLAB, we use the command **x** = inv (**A**)***B**. Similarly, for **XA = B**, we use the command **x** = **B***inv (**A**).

The basic computational unit in MATLAB is the matrix. A matrix expression is enclosed in square brackets, [ ]. Blanks or commas separate the column elements, and semicolons or carriage returns separate the rows.

```
>>A = [1  2   3   4 ;  5  6  7  8 ;  9  10  11  12]
A =
        1    2    3    4
        5    6    7    8
        9   10   11   12
```

The transpose of a simple matrix or a complex matrix is obtained by using the *apostrophe* key

```
>>B = A'
B =
        1    5    9
        2    6   10
        3    7   11
        4    8   12
```

Matrix multiplication is accomplished as follows:

```
>>C = A * B
C =
        30           70          110
        70          174          278
       110          278          446
>>C = B * A
C =
       107          122          137          152
       122          140          158          176
       137          158          179          200
       152          176          200          224
```

The inverse of a matrix D is obtained as

```
>>D = [1   2 ;  3   4]
D =
        1    2
        3    4
>>E = inv (D)
E =
      -2.0000    1.0000
       1.5000   -0.5000
```

Similarly, its eigenvalue is

```
>>eig (D)
ans =
        -0.3723
         5.3723
```

Matrix operations require that the matrix dimensions be compatible. If A is an $n \times m$ and B is a $p \times r$ then $A \pm B$ is allowed only if $n = p$ and $m = r$. Similarly, matrix product $A * B$ is allowed only if $m = p$.

Example 2.1. *Consider the two matrices:*

$$\mathbf{A} = \begin{bmatrix} 1 & 0 & 1 \\ 2 & 3 & 4 \\ -1 & 6 & 7 \end{bmatrix}$$

Using MATLAB, determine the following:

(a) $\mathbf{A} + \mathbf{B}$

(b) $\mathbf{AB}$

(c) $\mathbf{A}^2$

(d) $\mathbf{A}^T$

(e) $\mathbf{B}^{-1}$

(f) $\mathbf{B}^T\mathbf{A}^T$

(g) $\mathbf{A}^2 + \mathbf{B}^2 - \mathbf{AB}$

(h) *determinant of* $\mathbf{A}$, *determinant of* $\mathbf{B}$ *and determinant of* $\mathbf{AB}$.

Solution:

```
>> A= [1 0 1; 2 3 4; -1 6 7]
   A =
       1      0      1
       2      3      4
      -1      6      7
>> B= [7 4 2; 3 5 6; -1 2 1]
   B =
       7      4      2
       3      5      6
      -1      2      1
```

(a)
```
>> C=A+B
   C =
       8      4      3
       5      8     10
      -2      8      8
```

(b)
```
>>D=A*B
   D =
       6      6      3
      19     31     26
       4     40     41
```

(c)
```
>> E=A^2
   E =
       0      6      8
       4     33     42
       4     60     72
```

(*d*)   >> % Let F= transpose of A

        >> F=A'

        F =

           1   2   -1

           0   3   6

           1   4   7

(*e*)   >> H = inv (B)

        H =

           0.1111   0.0000   -0.2222

           0.1429   -0.1429   0.5714

         -0.1746   0.2857   -0.3651

(*f*)   >> J=B'*A'

        J =

           6   19   4

           6   31   40

           3   26   41

(*g*)   >> K= A^2 + B^2 -A * B

        K =

           53   52   45

           15   51   58

         -2   28   42

(*h*)   det (A) =12

        det (B) =-63

        det (A*B) =-756

Example 2.2. *Determine the eigenvalues and eigenvectors of A and B using MATLAB*

$$\mathbf{A} = \begin{bmatrix} 4 & 2 & -3 \\ -1 & 1 & 3 \\ 2 & 5 & 7 \end{bmatrix} \qquad \mathbf{B} = \begin{bmatrix} 1 & 2 & 3 \\ 8 & 7 & 6 \\ 5 & 3 & 1 \end{bmatrix}$$

Solution:

% Determine the eigenvalues and eigenvectors

A=[4 2 -3 ; -1 1 3 ; 2 5 7]

    A =

        4   2   -3

       -1   1   3

        2   5   7

eig(A)

    ans =

       0.5949

       3.0000

       8.4051

lamda=eig(A)

    lamda =

       0.5949

```
        3.0000
        8.4051
[V,D]=eig(A)
    V =
        -0.6713      0.9163      -0.3905
        0.6713      -0.3984       0.3905
        -0.3144      0.0398       0.8337
    D =
        0.5949       0            0
        0            3.0000       0
        0            0            8.4051
```

Example 2.3. *Determine the values of x, y, and z for the following set of linear algebraic* equations:

$$x_2 - 3x_3 = -5$$
$$2x_1 + 3x_2 - x_3 = 7$$
$$4x_1 + 5x_2 - 2x_3 = 10$$

Solution:

Here

$$\mathbf{A} = \begin{bmatrix} 0 & 1 & -3 \\ 2 & 3 & -1 \\ 4 & 5 & -2 \end{bmatrix} \quad \mathbf{B} = \begin{bmatrix} 5 \\ 7 \\ 10 \end{bmatrix} \text{ and } \mathbf{X} = \begin{bmatrix} x_1 \\ x_2 \\ x_3 \end{bmatrix}$$

$$\mathbf{AX} = \mathbf{B}$$
$$\mathbf{A^{-1}AX} = \mathbf{A^{-1}B}$$
$$\mathbf{IX} = \mathbf{A^{-1}B}$$

or

$$\mathbf{X} = \mathbf{A^{-1}B}$$

```
>> A = [0 1 -3; 2 3 -1; 4 5 -2];
>> B = [-5; 7; 10]
>> x = inv (A) * B
x =
        -1.0000
        4.0000
        3.0000
>> check = A * x
check =
              -5
               7
              10
%    Alternative method
>> x = A\B
x =
        -1
         4
         3
```

2.15 SCRIPT FILES

A *script* is a sequence of ordinary statements and functions used at the command prompt level. A script is invoked at the command prompt level by typing the file-name or by using the pull down menu. Scripts can also invoke other scripts.

The commands in the Command Window cannot be saved and executed again. Also, the Command Window is not interactive. To overcome these difficulties, the procedure is first to create a file with a list of commands, save it, and then run the file. In this way the commands contained are executed in the order they are listed when the file is run. In addition, as the need arises, one can change or modify the commands in the file, the file can be saved and run again. The files that are used in this fashion are known as *script files*. Thus, a script file is a text file that contains a sequence of MATLAB commands. Script file can be edited (corrected and/or changed) and executed many times.

2.15.1 CREATING AND SAVING A SCRIPT FILE

Any text editor can be used to create script files. In MATLAB script files are created and edited in the Editor/Debugger Window. This window can be opened from the Command Window. From the Command Window, select *File, New*, and then *M-file*. Once the window is open, the commands of the script file are typed line by line. The commands can also be typed in any text editor or word processor program and then copied and pasted in the Editor/Debugger Window. The second type of M-files is the *function file*. Function file enables the user to extend the basic library functions by adding one's own computational procedures. Function M-files are expected to return one or more results. Script files and function files may include reference to other MATLAB toolbox routines.

MATLAB function file begins with a header statement of the form:

```
function (name of result or results) = name (argument list)
```

Before a script file can be executed it must be saved. All script files must be saved with the extension ".*m*". MATLAB refers to them as m-files. When using MATLAB M-files editor, the files will automatically be saved with a ".*m*" extension. If any other text editor is used, the file must be saved with the ".*m*" extension, or MATLAB will not be able to find and run the script file. This is done by choosing *Save As...* from the *File* menu, selecting a location, and entering a name for the file. The names of user defined variables, predefined variables, MATLAB commands or functions should not be used to name script files.

2.15.2 RUNNING A SCRIPT FILE

A script file can be executed either by typing its name in the Command Window and then pressing the *Enter* key, directly from the Editor Window by clicking on the *Run* icon. The file is assumed to be in the current directory, or in the search path.

2.15.3 INPUT TO A SCRIPT FILE

There are three ways of assigning a value to a variable in a script file.

1. The variable is defined and assigned value in the script file.

2. The variable is defined and assigned value in the Command Window.

3. The variable is defined in the script file, but a specified value is entered in the Command Window when the script file is executed.

2.15.4 OUTPUT COMMANDS

There are two commands that are commonly used to generate output. They are the *disp* and *fprintf* commands.

1. The *disp* command

The *disp* command displays the elements of a variable without displaying the name of the variable, and displays text.

```
disp(name of a variable) or disp('text as string')
>>    A = [1 2 3 ; 4 5 6 ];
>> disp(A)
     1   2   3
     4   5   6
>> disp('Solution to the problem.')
     Solution to the problem.
```

2. The *fprintf* command

The *fprintf* command displays output (text and data) on the screen or saves it to a file. The output can be formatted using this command.

Example 2.4. *Write a function file Veccrossprod to compute the cross product of two vectors* **a**, *and* **b**, *where* $\mathbf{a} = (a_1, a_2, a_3)$, $b = (b_1, b_2, b_3)$, *and* $a \times b = (a_2 b_3 - a_3 b_2, a_3 b_1 - a_1 b_3, a_1 b_2 - a_2 b_1)$. *Verify the function by taking the cross products of pairs of unit vectors: (i, j), (j, k), etc.*

Solution:

```
function c = Veccrossprod (a, b);
% Veccrossprod : function to compute c = a x b where a and b are 3D vectors
% call syntax:
% c = Veccrossprod (a, b);
c = [a(2) * b(3)-a(3) * b(2); a(3) * b(1)-a(1) * b(3); a(1) * b(2)-a(2) * b(1)];
```

2.16 PROGRAMMING IN MATLAB

One most significant feature of MATLAB is its extendibility through user-written programs such as the M-files. M-files are ordinary ASCII text files written in MATLAB language. A function file is a subprogram.

2.16.1 RELATIONAL AND LOGICAL OPERATORS

A relational operator compares two numbers by finding whether a comparison statement is true or false. A logical operator examines true/false statements and produces a result which is true or false according to the specific operator. Relational and logical operators are used in mathematical expressions and also in combination with other commands, to make decision that control the flow a computer program.

MATLAB has six relational operators as shown in Table 2.21.

Table 2.21. Relational operators

Relational operator	Interpretation
<	Less than
<=	Less than or equal
>	Greater than
>=	Greater than or equal
= =	Equal
~ =	Not equal

The logical operators in MATLAB are shown in Table 2.22.

Table 2.22 Logical operators

Logical operator	Name	Description
& Example: $A\&B$	AND	Operates on two operands (A and B). If both are true, the result is true (1), otherwise the result is false (0).
| Example: $A\|B$	OR	Operates on two operands (A and B). If either one, or both are true, the result is true (1), otherwise (both are false) the result is false (0).
~ Example: $\sim A$	NOT	Operates on one operand (A). Gives the opposite of the operand. True (1) if the operand is false, and false (0) if the operand is true.

2.16.2 ORDER OF PRECEDENCE

The following Table 2.23 shows the order of precedence used by MATLAB.

Table 2.23

Precedence	Operation
1 (highest)	Parentheses (If nested parentheses exist, inner have precedence).
2	Exponentiation.
3	Logical NOT (~).
4	Multiplication, Division.
5	Addition, Subtraction.
6	Relational operators (>, <, >=, <=, = =, ~=).
7	Logical AND (&).
8 (lowest)	Logical OR (|).

2.16.3 BUILT-IN LOGICAL FUNCTIONS

The MATLAB built-in functions which are equivalent to the logical operators are:

and(A, B)   Equivalent to $A \& B$

or(A, B)   Equivalent to $A \mid B$

not(A)   Equivalent to $\sim A$

List the MATLAB logical built-in functions are described in Table 2.24.

Table 2.24 Additional logical built-in functions

Function	Description	Example
xor(a, b)	Exclusive or. Returns true (1) if one operand is true and the other is false	>>xor(8, − 1) ans = 0 >>xor(8, 0) ans = 1
all(A)	Returns 1 (true) if all elements in a vector A are true (nonzero). Returns 0 (false) if one or more elements are false (zero). If A is a matrix, treats columns of A as vectors, returns a vector with 1's and 0's.	>>A = [5 3 11 7 8 15] >>all(A) ans = 1 >>B = [3 6 11 4 0 13] >>all(B) ans = 0
any(A)	Returns 1 (true) if any element in a vector A is true (nonzero). Returns 0 (false) if all elements are false (zero). If A is a matrix, treats columns of A as vectors, returns a vector with 1's and 0's.	>>A = [5 0 14 0 0 13] >>any(A) ans = 1 >>B = [0 0 0 0 0 0] >>any(B) ans = 0
find(A) **find(A>d)**	If A is a vector, returns the indices of the nonzero elements. If A is a vector, returns the address of the elements that are larger than d (any relational operator can be used).	>>A = [0 7 4 2 8 0 0 3 9] >>find(A) ans = 2 3 4 5 8 9 >>find(A > 4) ans = 4 5 6

The truth table for the operation of the four logical operators, and, or, Xor, and not are summarized in Table 2.25.

Table 2.25 Truth table

INPUT		OUTPUT				
A	B	AND A&B	OR A|B	XOR (A,B)	NOT ~A	NOT ~B
false	false	false	false	false	true	true
false	true	false	true	true	true	false
true	false	false	true	true	false	true
true	true	true	true	false	false	false

2.16.4 CONDITIONAL STATEMENTS

A conditional statement is a command that allows MATLAB to make a decision of whether to execute a group of commands that follow the conditional statement or to skip these commands.

if conditional expression consists of relational and/or logical operators

```
if    a < 30
      count = count + 1
      disp a
      end
```

The general form of a simple **if** statement is as follows:

```
if    logical expression
            statements
      end
```

If the logical expression is true, the statements between the **if** statement and the **end** statement are executed. If the logical expression is false, then it goes to the statements following the **end** statement.

2.16.5 nested if STATEMENTS

Following is an example of *nested if* statements:

```
if    a < 30
      count = count + 1;
      disp(a);
      if    b > a
            b = 0;
      end
end
```

2.16.6 else AND elseif CLAUSES

The else clause allows to execute one set of statements if a logical expression is true and a different set if the logical expression is false.

```
%    variable name inc
     if   inc < 1
          x_inc = inc/10;
```

```
        else
        x_inc = 0.05;
        end
```

When several levels of **if-else** statements are nested, it may be difficult to find which logical expressions must be true (or false) to execute each set of statements. In such cases, the **elseif** clause is used to clarify the program logic.

2.16.7 MATLAB while STRUCTURES

There is a structure in MATLAB that combines the for loop with the features of the if block. This is called the *while loop* and has the form:

while *logical expression*

This set of statements is executed repeatedly as long as the logical expressions remain true (equals +1) or if the expression is a matrix rather than a simple scalar variable, as long as *all* the elements of the matrix remain nonzero.

end

In addition to the normal termination of a loop by means of the **end** statement, there are additional MATLAB commands available to interrupt the calculations. These commands are listed in Table 2.26 below:

Table 2.26

Command	Description
break	Terminates the execution of MATLAB **for** and **while** loops. In nested loops, **break** will terminate only the innermost loop in which it is placed.
return	Primarily used in MATLAB functions, **return** will cause a normal return from a function from the point at which the return statement is executed.
error (*'text'*)	Terminates execution and displays the message contained in text on the screen. Note, the text must be enclosed in single quotes.

The MATLAB functions used are summarized in Table 2.27 below:

Table 2.27

Function	Description
Relational operators	A MATLAB *logical relation* is a comparison between two variables **x** and **y** of the same size effected by one of the six operators, <, <=, >, >=, = =, ~=. The comparison involves corresponding elements of **x** and **y**, and yields a matrix or scalar of the same size with values of "true" or "false" for each of its elements. In MATLAB, the value of "false" is zero, and "true" has a value of one. Any nonzero quantity is interpreted as "true".
Combinatorial operators	The operators **&** (AND) and **|** (OR) may be used to combine two logical expressions.
all, any	If **x** is a vector, **all(x)** returns a value of one if *all* of the elements of **x** are nonzero, and a value of zero otherwise. When **X** is a matrix, all(**X**) returns a row vector of ones or zeros obtained by applying all to each of the columns of **X**. The function **any** operates similarly if any of the elements of **x** are nonzero.

Function	*Description*

find	If **x** is a vector, **i** = **find(x)** returns the indices of those elements of **x** that are nonzero (*i.e.*, true). Thus, replacing all the negative elements of **x** by zero could be accomplished by
	
	  i = find(x < 0);
	  x(i) = zeros(size(i));
	
	If **X** is a matrix, **[i,j]** = **find(X)** operates similarly and returns the row-column indices of nonzero elements.

if, else, elseif	The several forms of MATLAB **if** blocks are as follows: **if** *variable* **if** *variable* 1 **if** *variable* 1 block of statements block of statements block of statements executed if *variable* executed if *variable* 1 executed if *variable* 1 is "true", *i.e.*, nonzero is "true", *i.e.*, nonzero is "true", **end** **else** **elseif** *variable* 2 block of statements block of statements executed if *variable l* executed if *variable* 2 is "false", i.e., zero is "true", **end** **else** **end** block of statements executed if neither *variable* is "true"
break	Terminates the execution of a **for** or **while** loop. Only the innermost loop in which **break** is encountered will be terminated.
return	Causes the function to return at that point to the calling routine. MATLAB *M-file* functions will return normally without this statement.
error (*'text'*)	Within a loop or function, if the statement **error(***'text'***)** is encountered, the loop or function is terminated, and the text is displayed.

while	The form of the MATLAB **while** loop is
	
	  while *variable*
	
	    block of statements executed as long as the value of *variable* is
	      "true"; *i.e.*, nonzero
	
	  end
	
	Useful when a function *F* itself calls a second "dummy" function "*f*". For example, the function *F* might find the root of an arbitrary function identified as a generic *f(x)*. Then, the name of the actual *M-file* function, say **fname**, is passed as a *character string* to the function *F* either through its argument list or as a global variable, and the function is evaluated within *F* by means of **feval**. The use of **feval(name, x1, x2, ..., xn)**, where **fname** is a variable containing the name of the function as a character string; *i.e.*, enclosed in single quotes, and **x1, x2, ..., xn** are the variables needed in the argument list of function **fname**.

2.17 GRAPHICS

MATLAB has many commands that can be used to create basic 2-D plots, overlay plots, specialized 2-D plots, 3-D plots, mesh, and surface plots.

2.17.1 BASIC 2-D PLOTS

The basic command for producing a simple 2-D plot is

plot(*x* values, *y* values, 'style option')

where   *x* values and *y* values are vectors containing the *x*- and *y*-coordinates of points on the graph.

style   option is an optional argument that specifies the color, line-style, and the point-marker style.

The style option in the plot command is a character string that consists of 1, 2, or 3 characters that specify the color and/or the line style. The different color, line-style and marker-style options are summarized in Table 2.28.

Table 2.28 Color, line-style, and marker-style options

Color style-option		Line style-option		Marker style-option	
y	yellow	–	solid	+	plus sign
m	magenta	– –	dashes	O	circle
c	cyan	:	dotted	*	asterisk
r	red	–.	dash-dot	*x*	x-mark
g	green			.	point
b	blue			^	up triangle
w	white			*s*	square
k	black			*d*	diamond, etc.

2.17.2 SPECIALIZED 2-D PLOTS

There are several specialized graphics functions available in MATLAB for 2-D plots. The list of functions commonly used in MATLAB for plotting *x-y* data are given in Table 2.29.

Table 2.29 List of functions for plotting x-y data

Function	Description
area	Creates a filled area plot.
bar	Creates a bar graph.
barh	Creates a horizontal bar graph.
comet	Makes an animated 2-*D* plot.
compass	Creates arrow graph for complex numbers.
contour	Makes contour plots.
contourf	Makes filled contour plots.
errorbar	Plots a graph and puts error bars.
feather	Makes a feather plot.
fill	Draws filled polygons of specified color.
fplot	Plots a function of a single variable.
hist	Makes histograms.
loglog	Creates plot with log scale on both *x* and *y* axes.
pareto	Makes pareto plots.
pcolor	Makes pseudo color plot of matrix.

Command	Description
pie	Creates a pie chart.
plotyy	Makes a double y-axis plot.
plotmatrix	Makes a scatter plot of a matrix.
polar	Plots curves in polar coordinates.
quiver	Plots vector fields.
rose	Makes angled histograms.
scatter	Creates a scatter plot.
semilogx	Makes semilog plot with log scale on the x-axis.
semilogy	Makes semilog plot with log scale on the y-axis.
stairs	Plots a stair graph.
stem	Plots a stem graph.

2.17.2.1 Overlay Plots

There are three ways of generating overlay plots in MATLAB, they are:

(a) Plot command

(b) Hold command

(c) Line command

(a) **Plot command.** Example 2.5(a) shows the use of plot command used with matrix argument, each column of the second argument matrix plotted against the corresponding column of the first argument matrix.

(b) **Hold command.** Invoking hold on at any point during a session freezes the current plot in the graphics window. All the next plots generated by the plot command are added to the exiting plot. See Example 2.5(a).

(c) **Line command.** The line command takes a pair of vectors (or a triplet in 3-D) followed by a parameter name/parameter value pairs as argument. For instance, the command: *line* (*x data, y data, parameter name, parameter value*) adds lines to the existing axes. See Example E2.5(a).

2.17.3 3-D PLOTS

MATLAB provides various options for displaying three-dimensional data. They include line and wire, surface, mesh plots, among many others. More information can be found in the Help Window under Plotting and Data visualization. Table 2.30 lists commonly used functions.

Table 2.30 Functions used for 3-D graphics

Command	Description
plot3	Plots three-dimensional graph of the trajectory of a set of three parametric equations $x(t)$, $y(t)$, and $z(t)$ can be obtained using **plot3(x,y,z).**
meshgrid	If **x** and **y** are two vectors containing a range of points for the evaluation of a function, **[X,Y]** = **meshgrid(x, y)** returns two rectangular matrices containing the x and y values at each point of a two-dimensional grid.
mesh(X,Y,z)	If **X** and **Y** are rectangular arrays containing the values of the x and y coordinates at each point of a rectangular grid , and if z is the value of a function evaluated at each of these points, **mesh(X,Y,z)** will produce a three-dimensional perspective graph of the points. The same results can be obtained with **mesh(x,y,z)**.

Command	Description
meshc, meshz	If the xy grid is rectangular, these two functions are merely variations of the basic plotting program **mesh,** and they operate in an identical fashion. **meshc** will produce a corresponding contour plot drawn on the xy plane below the three-dimensional figure, and **meshz** will add a vertical wall to the outside features of the figures drawn by **mesh.**
surf	Produces a three-dimensional perspective drawing. Its use is usually to draw surfaces, as opposed to plotting functions, although the actual tasks are quite similar. The output of **surf** will be a *shaded* figure. If row vectors of length n are defined by $x = r \cos \theta$ and $y = r \sin \theta$, with $0 \le \theta \le 2\pi$, they correspond to a circle of radius r. If $\vec{r}$ is a *column* vector equal to **r = [0 1 2]'**; then **z = r*ones(size(x))** will be a rectangular, $3 \times n$, arrays of 0's and 2's, and **surf(x, y, z)** will produce a shaded surface bounded by three circles; *i.e.*, a cone.
surfc	This function is related to **surf** in the same way that **meshc** is related to mesh.
colormap	Used to change the default coloring of a figure. See the MATLAB reference manual or the help file.
shading	Controls the type of color shading used in drawing figures. See the MATLAB reference manual or the help file.
view	**view(az,el)** controls the perspective view of a three-dimensional plot. The view of the figure is from angle "**el**" above the xy plane with the coordinate axes (and the figure) rotated by an angle "**az**" in a clockwise direction about the z axis. Both angles are in degrees. The default values are **az** $= 37\frac{1}{2}°$ and **el** $= 30°$.
axis	Determines or changes the scaling of a plot. If the coordinate axis limits of a two-dimensional or three-dimensional graph are contained in the row vector $r = [x_{min}, x_{max}, y_{min}, y_{max}, z_{min}, z_{max}]$, **axis** will return the values in this vector, and **axis(r)** can be used to alter them. The coordinate axes can be turned *on* and *off* with **axis('on')** and **axis('off')**. A few other string constant inputs to **axis** and their effects are given below:
	axis('equal') x and y scaling are forced to be the same.
	axis('square') The box formed by the axes is square.
	axis('auto') Restores the scaling to default settings.
	axis('normal') Restoring the scaling to full size, removing any effects of **square** or **equal** settings.
	axis ('image') Alters the aspect ratio and the scaling so the screen pixels are square shaped rather than rectangular.
contour	The use is **contour(x,y,z)**. A default value of $N = 10$ contour lines will be drawn. An optional fourth argument can be used to control the number of contour lines that are drawn. **contour(x,y,z,N)**, if **N** is a positive integer, will draw **N** contour lines, and **contour(x,y,z,V)**, if **V** is a vector containing values in the range of z values, will draw contour lines at each value of $z = V$.
plot3	Plots lines or curves in three dimensions. If **x, y,** and **z** are vectors of equal length, **plot3(x,y,z)** will draw, on a three-dimensional coordinate axis system, the lines connecting the points. A fourth argument, representing the color and symbols to be used at each point, can be added in exactly the same manner as with **plot.**
grid	**grid on** adds grid lines to a two-dimensional or three-dimensional graph; **grid off** removes them.
slice	Draws "slices" of a volume at a particular location within the volume.

Example 2.5. *(a) Generate an overlay plot for plotting three lines*

$$y_1 = \sin t$$
$$y_2 = t$$

$$y_3 = t - \frac{t^3}{3!} + \frac{t^5}{5!} + \frac{t^7}{7!}$$

Use (i) the plot command
(ii) the hold command
(iii) the line command

(b) Use the functions for plotting x-y data for plotting the following functions.
  (i) f(t) = t cost
  $0 \le t \le 10\pi$
 (ii) x = e^t
  $y = 100 + e^{3t}$
  $0 \le t \le 2\pi.$

Solution:

(a) overlay plot

(i)

```
% using the plot command
t = linspace(0, 2*pi, 100);
y1 = sin(t); y2 = t;
y3 = t - (t.^3)/6 + (t.^5)/120 - (t.^7)/5040;
plot(t, y1, t, y2, '-', t, y3, 'o')
axis([0 5 -1 5])
xlabel('t')
ylabel('sin(t) approximation')
title('sin(t) function')
text(3.5,0, 'sin(t)')
gtext('Linear approximation')
gtext('4-term approximation')
```

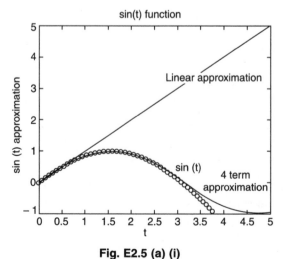

Fig. E2.5 (a) (i)

(*ii*)

```
% using the hold command
x = linspace(0, 2*pi, 100); y1 = sin(x);
plot(x, y1)
hold on
y2 = x; plot(x, y2, '-' )
y3 = x - (x.^3)/6 + (x.^5)/120 - (t.^7)/5040;
plot(x, y3, 'o')
axis([0 5 -1 5])
hold off
```

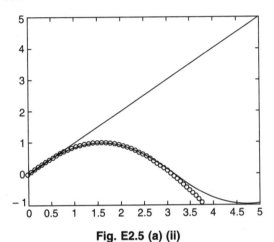

Fig. E2.5 (a) (ii)

(*iii*) % using the line command

```
t = linspace(0, 2*pi, 100);
y1 = sin(t);
y2 = t;
y3 = t - (t.^3)/6 + (t.^5)/120 - (t.^7)/5040;
plot(t, y1)
line(t, y2, 'linestyle', '-')
line(t, y3, 'marker', 'o')
axis([0 5 -1 5])
xlabel('t')
ylabel('sin(t) approximation')
title('sin(t) function')
legend('sin(t)', 'linear approx', '7th order approx')
```

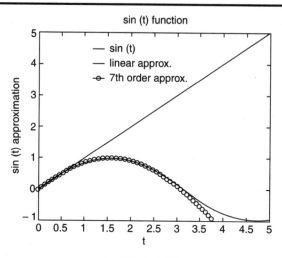

Fig. E2.5(a) (iii)

(*b*) Using Table 2.29 functions

(*i*)   `fplot('x.*cos(x)', [0 10*pi])`

This will give the following figure (Fig. E2.5 (*b*) (*i*))

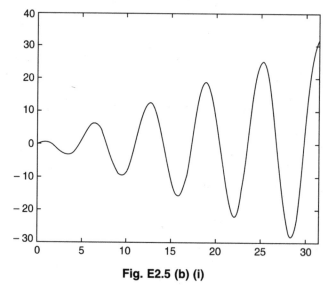

Fig. E2.5 (b) (i)

(*ii*)
```
t = linspace(0, 2*pi, 200);
x = exp(t);
y = 100 + exp(3*t);
loglog(x, y), grid
```

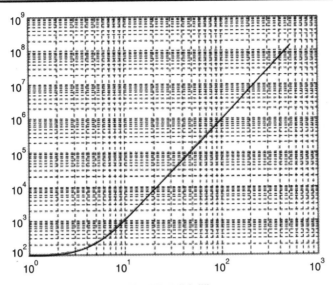

Fig. E2.5 (b) (ii)

Example 2.6. (*a*) *Plot the parametric space curve of*

$$x(t) = t$$
$$y(t) = t^2$$
$$z(t) = t^3 \qquad\qquad 0 \le t \le 2.0$$

(*b*) $z = -7/(1 + x^2 + y^2)$ $\qquad |x| \le 5, |y| \le 5$

Solution:

(*a*)
```
>> t=linspace(0, 2,100);
>> x=t; y=t. ^2; z=t. ^3;
>> plot3(x, y, z), grid
```
The plot is shown in Figure E2.6 (*a*).

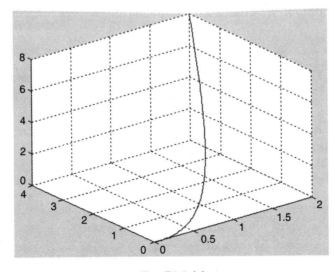

Fig. E2.6 (a)

```
(b)   >> t=linspace(0, 2,100);
      >> x=t; y=t. ^2; z=t. ^3;
      >> plot3(x, y, z), grid
      >> t=linspace(-5,5,50);y=x;
      >> z=-7./(1+x.^2+y.^2);
      >> mesh(z)
```
The plot is shown in Figure E2.6(b).

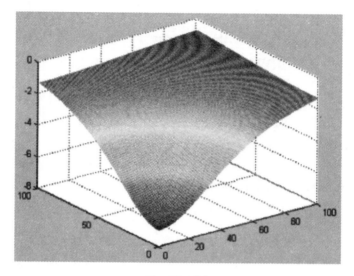

Fig. E2.6(b)

2.17.4 SAVING AND PRINTING GRAPHS

To obtain a hardcopy of a graph, type *print* in the Command Window after the graph appears in the Figure Window. The figure can also be saved into a specified file in the PostScripter or Encapsulated PostScript (EPS) format. The command to save graphics to a file is

$$\textbf{print} - d \text{ devicetype} - \text{options filename}$$

where device type for PostScript printers are listed in the following Table 2.31.

Table 2.31 Devicetype for Post Script printers

Devicetype	Description	Devicetype	Description
ps	Black and white PostScript	**eps**	Black and white EPSF
psc	Color PostScript	**epsc**	Color EPSF
ps2	Level 2 BW PostScript	**eps2**	Level 2 black and white EPSF
psc2	Level 2 color PostScript	**epsc2**	Level 2 color EPSF

MATLAB can also generate a graphics file in the following popular formats among others.

–dill   saves file in Adobe Illustrator format.

–djpeg   saves file as a JPEG image.

–dtiff   saves file as a compressed TIFF image.

–dmfile   saves file as an M-file with graphics handles.

2.18 INPUT/OUTPUT IN MATLAB

In this section, we present some of the many available commands in MATLAB for reading data from an external file into a MATLAB matrix, or writing the numbers computed in MATLAB into such an external file.

2.18.1 THE fopen STATEMENT

To have the MATLAB read or write a separate data file of numerical values, we need to *connect* the file to the executing MATLAB program. The MATLAB functions used are summarized in Table 2.32.

Table 2.32 MATLAB functions used for input/output

Function	*Description*

fopen	Connects an existing file to MATLAB or to create a new file from MATLAB.
	
	    fid = **fopen**('*Filename*', *permission code*);
	
	where, if **fopen** is successful, **fid** will be returned as a positive integer greater than 2. When unsuccessful, a value of –1 is returned. Both the file name and the permission code are string constants enclosed in single quotes. The permission code can be a variety of flags that specify whether or not the file can be written to, read from, appended to, or a combination of these. Some common codes are:
	
	    Code *Meaning*
	    'r' read only
	    'w' write only
	    'r+' read and write
	    'a+' read and append
	The **fopen** statement positions the file at the beginning.

fclose	Disconnects a file from the operating MATLAB program. The use is **fclose(fid)**, where **fid** is the *file identification number* of the file returned by **fopen.fclose('all')** will close all files.
fscanf	Reads opened files. The use is **A = fscanf(fid**, *FORMAT, SIZE*) where *FORMAT* specifies the types of numbers (integers, reals with or without exponent, character strings) and their arrangement in the data file, and optional *SIZE* determines how many quantities are to be read and how they are to be arranged into the matrix **A**. If *SIZE* is omitted, the entire file is read. The *FORMAT* field is a string (enclosed in single quotes) specifying the form of the numbers in the file. The *type* of each number is characterized by a percent sign (%), followed by a letter (**i** or **d** for integers, **e** or **f** for floating-point numbers with or without exponents). Between the percent sign and the type code, one can insert an integer specifying the maximum width of the field.

Function	Description
fprintf	Writes files previously opened. **fprintf(fid,** *FORMAT,* **A)** where **fid** and *FORMAT* have the same meaning as for **fscanf**, with the exception that for output formats the string must end with **\n**, designating the end of a line of output.

2.19   SYMBOLIC MATHEMATICS

In Secs. 2.1 to 2.18, the capability of MATLAB for numerical computations have been described. In this section some of MATLAB's capabilities for symbolic manipulations will be presented. Specifically, the symbolic expressions, symbolic algebra, simplification of mathematical expressions, operations on symbolic expressions, solution of a single equation or a set of linear algebraic equations, solutions to differential equations, differentiation and integration of functions using MATLAB are presented.

2.19.1 SYMBOLIC EXPRESSIONS

A symbolic expression is stored in MATLAB as a *character string*. A single quote marks are used to define the symbolic expression. For instance:

$$\text{`sin}(y/x)\text{'; `}x^4 + 5*x^3 + 7*x^2 - 7\text{'}$$

The independent variable in many functions is specified as an additional function argument. If an independent variable is not specified, then MATLAB will pick one. When several variables exist, MATLAB will pick the one that is a single lower case letter (except i and j), which is closest to x alphabetically.

The independent variable is returned by the function *symvar*,

symvar(s)   Returns the independent variable for the symbolic expression s.

For example:

Expression s	**symvar(s)**
'5 * c * d + 34'	d
'sin(y/x)'	x

In MATLAB, a number of functions are available to simplify mathematical expressions by expanding the terms, factoring expressions, collecting coefficients, or simplifying the expression. For instance:

expand(s)   Performs an expansion of s.

A summary of these expressions is given in Table 2.33. A summary of basic operations is given in Table 2.34. The standard arithmetic operation (Table 2.35) is applied to symbolic expressions using symbolic functions. These symbolic expressions are summarized in Table 2.36.

Table 2.33

Simplification	
collect	Collect common terms
expand	Expand polynomials and elementary functions
factor	Factorization
horner	Nested polynomial representation
numden	Numerator and denominator
simple	Search for shortest form
simplify	Simplification
subexpr	Rewrite in terms of subexpressions

Table 2.34

Basic Operations	
ccode	C code representation of a symbolic expression
conj	Complex conjugate
findsym	Determine symbolic variables
fortran	Fortran representation of a symbolic expression
imag	Imaginary part of a complex number
latex	LaTeX representation of a symbolic expression
pretty	Pretty prints a symbolic expression
real	Real part of an imaginary number
sym	Create symbolic object
syms	Shortcut for creating multiple symbolic objects

Table 2.35

Arithmetic Operations	
+	Addition
−	Subtraction
*	Multiplication
.*	Array multiplication
/	Right division
./	Array right division
\	Left division
.\	Array left division
^	Matrix or scalar raised to a power
.^	Array raised to a power
'	Complex conjugate transpose
.'	Real transpose

Table 2.36

	Symbolic expressions
horner(S)	Transposes **S** into its Horner, or nested, representation.
numden(S)	Returns two symbolic expressions that represent, respectively, the numerator expression and the denominator expression for the rational representation of **S**.
numeric(S)	Converts **S** to a numeric form (**S** must not contain any symbolic variables).
poly2sym(c)	Converts a polynomial coefficient vector **c** to a symbolic polynomial.
pretty(S)	Prints **S** in an output form that resembles typeset mathematics.
sym2poly(S)	Converts **S** to a polynomial coefficient vector.*
symadd(A,B)	Performs a symbolic addition, **A + B**.
symdiv(A,B)	Performs a symbolic division, **A/B**.
symmul(A,B)	Performs a symbolic multiplication, **A * B**.
sympow(S,p)	Performs a symbolic power, **S^p**.
symsub(A,B)	Performs a symbolic subtraction, **A – B**.

2.19.2 SOLUTION TO DIFFERENTIAL EQUATIONS

Symbolic math functions can be used to solve a single equation, a system of equations, and differential equations. For example:

solve(f)   Solves a symbolic equation **f** for its symbolic variable. If **f** is a symbolic expression, this function solves the equation **f** = 0 for its symbolic variable.

solve(f1, ... fn)   Solves the system of equations represented by **f1, ..., fn.**

The symbolic function for solving ordinary differential equation is **dsolve** as shown below:

dsolve('equation', 'condition')   Symbolically solves the ordinary differential equation specified by '**equation**'. The optional argument '**condition**' specifies a boundary or initial condition.

The symbolic equation uses the letter **D** to denote differentiation with respect to the independent variable. A **D** followed by a digit denotes repeated differentiation. Thus, **Dy** represents dy/dx, and **D2y** represents d^2y/dx^2. For example, given the ordinary second order differential equation;

$$\frac{d^2x}{dt^2} + 5\frac{dx}{dt} + 3x = 7$$

with the initial conditions $x(0) = 0$ and $\dot{x}(0) = 1$.

The MATLAB statement that determine the symbolic solution for the above differential equation is the following:

```
x = dsolve('D2x = -5*Dx-3*x+7', 'x(0)=0', 'Dx(0)=1')
```

The symbolic functions are summarized in Table 2.37.

Table 2.37

Solution of Equations	
compose	Functional composition
dsolve	Solution of differential equations
finverse	Functional inverse
solve	Solution of algebraic equations

2.19.3 CALCULUS

There are four forms by which the symbolic derivative of a symbolic expression is obtained in MATLAB. They are:

diff(f) — Returns the derivative of the expression **f** with respect to the default independent variable.

diff(f, 't') — Returns the derivative of the expression **f** with respect to the variable t.

diff(f,n) — Returns the nth derivative of the expression **f** with respect to the default independent variable.

diff(f, 't',n) — Returns the nth derivative of the expression **f** with respect to the variable t.

The various forms that are used in MATLAB to find the integral of a symbolic expression **f** are given below and summarized in Table 2.38.

int(f) — Returns the integral of the expression **f** with respect to the default independent variable.

int(f, 't') — Returns the integral of the expression **f** with respect to the variable t.

int(f,a,b) — Returns the integral of the expression **f** with respect to the default independent variable evaluated over the interval **[a,b]**, where **a** and **b** are numeric expressions.

int(f, 't',a,b) — Returns the integral of the expression **f** with respect to the variable **t** evaluated over the interval **[a,b]**, where **a** and **b** are numeric expressions.

int(f, 'm', 'n') — Returns the integral of the expression **f** with respect to the default independent variable evaluated over the interval **[m,n]**, where m and n are numeric expressions.

The other symbolic functions for pedagogical and graphical applications, conversions, integral transforms, and linear algebra are summarized in Tables 2.38 to 2.42.

Table 2.38

Calculus	
diff	Differentiate
int	Integrate
jacobian	Jacobian matrix
limit	Limit of an expression
symsum	Summation of series
taylor	Taylor series expansion

Table 2.39

	Pedagogical and Graphical Applications
ezcontour	Contour plotter
ezcontourf	Filled contour plotter
ezmesh	Mesh plotter
ezmeshc	Combined mesh and contour plotter
ezplot	Function plotter
ezplot	Easy-to-use function plotter
ezplot3	Three-dimensional curve plotter
ezpolar	Polar coordinate plotter
ezsurf	Surface plotter
ezsurfc	Combined surface and contour plotter
funtool	Function calculator
rsums	Riemann sums
taylortool	Taylor series calculator

Table 2.40

	Conversions
char	Convert sym object to string
double	Convert symbolic matrix to double
poly2sym	Function calculator
sym2poly	Symbolic polynomial to coefficient vector

Table 2.41

	Integral Transforms
fourier	Fourier transform
ifourier	Inverse Fourier transform
ilaplace	Inverse Laplace transform
iztrans	Inverse Z-transform
laplace	Laplace transform
ztrans	Z-transform

Table 2.42

	Linear Algebra
colspace	Basis for column space
det	Determinant
diag	Create or extract diagonals
eig	Eigenvalues and eigenvectors
expm	Matrix exponential
inv	Matrix inverse
jordan	Jordan canonical form
null	Basis for null space
poly	Characteristic polynomial
rank	Matrix rank
rref	Reduced row echelon form
svd	Singular value decomposition
tril	Lower triangle
triu	Upper triangle

2.20 THE LAPLACE TRANSFORMS

The Laplace transformation method is an operational method that can be used to find the transforms of time functions, the inverse Laplace transformation using the partial-fraction expansion of $B(s)/A(s)$, where $A(s)$ and $B(s)$ are polynomials in s. In this Chapter, we present the computational methods with MATLAB to obtain the partial-fraction expansion of $B(s)/A(s)$ and the zeros and poles of $B(s)/A(s)$.

MATLAB can be used to obtain the partial-fraction expansion of the ratio of two polynomials, $B(s)/A(s)$ as follows:

$$\frac{B(s)}{A(s)} = \frac{num}{den} = \frac{b(1)s^n + b(2)s^{n-1} + ... + b(n)}{a(1)s^n + a(2)s^{n-1} + ... + a(n)}$$

where $a(1) \neq 0$ and num and den are row vectors. The coefficients of the numerator and denominator of $B(s)/A(s)$ are specified by the num and den vectors.

Hence $num = [b(1) \quad b(2) \quad ... \quad b(n)]$

$den = [a(1) \quad a(2) \quad ... \quad a(n)]$

The MATLAB command

r, p, k = residue(num, den)

is used to determine the residues, poles, and direct terms of a partial-fraction expansion of the ratio of two polynomials $B(s)$ and $A(s)$ is then given by

$$\frac{B(s)}{A(s)} = k(s) + \frac{r(1)}{s - p(1)} + \frac{r(2)}{s - p(1)} + ... + \frac{r(n)}{s - p(n)}$$

The MATLAB command [**num, den**] = **residue(r, p, k)** where r, p, k are the output from MATLAB converts the partial fraction expansion back to the polynomial ratio $B(s)/A(s)$.

The command **printsys (num,den's')** prints the num/den in terms of the ratio of polynomials in s.

The command **ilaplace** will find the inverse Laplace transform of a Laplace function.

2.20.1 FINDING ZEROS AND POLES OF B(s)/A(s)

The MATLAB command [**z,p,k**] = **tf2zp(num,den)** is used to find the zeros, poles, and gain K of $B(s)/A(s)$.

If the zeros, poles, and gain K are given, the following MATLAB command can be used to find the original num/den:

$$[\textbf{num,den}] = \textbf{zp2tf (z,p,k)}$$

2.21 CONTROL SYSTEMS

MATLAB has an extensive set of functions for the analysis and design of control systems. They involve matrix operations, root determination, model conversions, and plotting of complex functions. These functions are found in MATLAB's control systems toolbox. The analytical techniques used by MATLAB for the analysis and design of control systems assume the processes that are linear and time invariant. MATLAB uses models in the form of *transfer-functions or state-space equations*.

2.21.1 TRANSFER FUNCTIONS

The transfer function of a linear time invariant system is expressed as a ratio of two polynomials. The transfer function for a single input and a single output (SISO) system is written as

$$H(s) = \frac{b_0 s^n + b_1 s^{n-1} + \ldots b_{n-1}s + b_n}{a_0 s^m + a_1 s^{m-1} + \ldots + a_{m-1}s + a_m}$$

when the numerator and denominator of a transfer function are factored into the *zero-pole-gain form*, it is given by

$$H(s) = k\frac{(s - z_1)(s - z_2)\ldots(s - z_n)}{(s - p_1)(s - p_2)\ldots(s - p_m)}$$

The *state-space model* representation of a linear control system s is written as

$$\dot{x} = Ax + Bu$$
$$y = Cx + Du$$

2.21.2 MODEL CONVERSION

There are a number of functions in MATLAB that can be used to convert from one model to another. These conversion functions and their applications are summarized in Table 2.43.

Table 2.43 Model conversion functions

Function	Purpose
c2d	Continuous state-space to discrete state-space
residue	Partial-fraction expansion
ss3tf	State-space to transfer function
ss2zp	State-space to zero-pole-gain
tf2ss	Transfer function to state-space
tf2zp	Transfer function to zero-pole-gain
zp2ss	Zero-pole-gain to state-space
zp2tf	Zero-pole-gain to transfer function

Residue Function: The **residue** function converts the polynomial transfer function

$$H(s) = \frac{b_0 s^n + b_1 s^{n-1} + \ldots + b_{n-1}s + b_n}{a_0 s^m + a_1 s^{m-1} + \ldots + a_{m-1}s + a_m}$$

to the partial fraction transfer function

$$H(s) = \frac{r_1}{s - p_1} + \frac{r_2}{s - p_2} + \ldots + \frac{r_n}{s - p_n} + k(s)$$

[r,p,k] = residue(B, A) Determine the vectors **r, p,** and **k,** which contain the residue values, the poles, and the direct terms from the partial-fraction expansion. The inputs are the polynomial coefficients **B** and **A** from the numerator and denominator of the transfer function, respectively.

ss2tf Function: The **ss2tf** function converts the continuous-time, state-space equations

$$x' = Ax + Bu$$
$$y = Cx + Du$$

to the polynomial transfer function

$$H(s) = \frac{b_0 s^n + b_1 s^{n-1} + \ldots + b_{n-1}s + b_n}{a_0 s^m + a_1 s^{m-1} + \ldots + a_{m-1}s + a_m}$$

The function has two output matrices:

[num,den] = ss2tf(A,B,C,D,iu) Computes vectors **num** and **den** containing the coefficients, in descending powers of **s,** of the numerator and denominator of the polynomial transfer function for the **iu**th input. The input arguments **A,B,C,** and **D** are the matrices of the state-space equations corresponding to the **iu**th input, where **iu** is the number of the input for a multi-input system. In the case of a single-input system, **iu** is 1.

ss2zp Function: The **ss2zp** function converts the continuous-time, state-space equations

$$x' = Ax + Bu$$
$$y = Cx + Du$$

to the zero-pole-gain transfer function

$$H(s) = k \frac{(s - z_1)(s - z_2)\dots(s - z_n)}{(s - p_1)(s - p_2)\dots(s - p_m)}$$

The function has three output matrices:

[z,p,k] = ss2zp(A,B,C,D,iu) Determines the zeros (**z**) and poles (**p**) of the zero-pole-gain transfer function for the iu^{th} input, along with the associated gain (**k**). The input matrices **A, B, C,** and **D** of the state-space equations correspond to the iu^{th} input, where **iu** is the number of the input for a multi-input system. In the case of a single-input system **iu** is 1.

tf2ss Function: The **ts2ss** function converts the polynomial transfer function

$$H(s) = \frac{b_0 s^n + b_1 s^{n-1} + \dots + b_{n-1}s + b_n}{a_0 s^m + a_1 s^{m-1} + \dots + a_{m-1}s + a_m}$$

to the controller-canonical form state-space equations

$$x' = Ax + Bu$$
$$y = Cx + Du$$

The function has four output matrices:

[A,B,C,D] = tf2ss(num,den) Determines the matrices **A, B, C, and D** of the controller-canonical form state-space equations. The input arguments num and den contain the coefficients, in descending powers of **s**, of the numerator and denominator polynomials of the transfer function that is to be converted.

tf2zp Function: The **tf2zp** function converts the polynomial transfer function

$$H(s) = \frac{b_0 s^n + b_1 s^{n-1} + \dots + b_{n-1}s + b_n}{a_0 s^m + a_1 s^{m-1} + \dots + a_{m-1}s + a_m}$$

to the zero-pole-gain transfer function

$$H(s) = k \frac{(s - z_1)(s - z_2)\dots(s - z_n)}{(s - p_1)(s - p_2)\dots(s - p_m)}$$

The function has three output matrices:

[z,p,k] = tf2zp(num,den) Determines the zeros (**z**), poles (**p**) and associated gain (**k**) of the zero-pole-gain transfer function using the coefficients, in descending powers of **s**, of the numerator and denominator of the polynomial transfer function that is to be converted.

zp2tf Function: The **zp2tf** function converts the zero-pole-gain transfer function

$$H(s) = k \frac{(s - z_1)(s - z_2)\dots(s - z_n)}{(s - p_1)(s - p_2)\dots(s - p_m)}$$

to the polynomial transfer function

$$H(s) = \frac{b_0 s^n + b_1 s^{n-1} + \ldots + b_{n-1} s + b_n}{a_0 s^m + a_1 s^{m-1} + \ldots + a_{m-1} s + a_m}$$

The function has two output matrices:

[num,den] = zp2tf(z,p,k) Determines the vectors num and den containing the coefficients, in descending powers of **s**, of the numerator and denominator of the polynomial transfer function. **p** is a column vector of the pole locations of the zero-pole-gain transfer function, **z** is a matrix of the corresponding zero locations, having one column for each output of a multi-output system, **k** is the gain of the zero-pole-gain transfer function. In the case of a single-output system, **z** is a column vector of the zero locations corresponding to the pole locations of vector **p**.

zp2ss Function: The **zp2ss** function converts the zero-pole-gain transfer function

$$H(s) = k \frac{(s - z_1)(s - z_2) \ldots (s - z_n)}{(s - p_1)(s - p_2) \ldots (s - p_m)}$$

to the controller-canonical form state-space equations

$$x' = Ax + Bu$$
$$y = Cx + Du$$

The function has four output matrices:

[A,B,C,D] = zp2ss(z,p,k) Determines the matrices **A, B, C,** and **D** of the control-canonical form state-space equations. **p** is a column vector of the pole locations of the zero-pole-gain transfer function, **z** is a matrix of the corresponding zero locations, having one column for each output of a multi-output system, **k** is the gain of the zero-pole-gain transfer function. In the case of a single-output system, **z** is a column vector of the zero locations corresponding to the pole locations of vector **p**.

2.22 THE LAPLACE TRANSFORMS

MATLAB can be used to obtain the partial-fraction expansion of the ratio of two polynomials, $B(s)/A(s)$ as follows:

$$\frac{B(s)}{A(s)} = \frac{num}{den} = \frac{b(1)s^n + b(2)s^{n-1} + \ldots + b(n)}{a(1)s^n + a(2)s^{n-1} + \ldots + a(n)}$$

where $a(1) \neq 0$ and num and den are row vectors. The coefficients of the numerator and denominator of $B(s)/A(s)$ are specified by the num and den vectors.

Hence $num = [b(1) \quad b(2) \quad \ldots \quad b(n)]$

$den = [a(1) \quad a(2) \quad \ldots \quad a(n)]$

The MATLAB command

r, p, k = residue(num, den)

is used to determine the residues, poles, and direct terms of a partial-fraction expansion of the ratio of two polynomials $B(s)$ and $A(s)$ is then given by

$$\frac{B(s)}{A(s)} = k(s) + \frac{r(1)}{s - p(1)} + \frac{r(2)}{s - p(2)} + \ldots + \frac{r(n)}{s - p(n)}$$

The MATLAB command **[num, den] = residue(r, p, k)** where r, p, k are the output from MATLAB converts the partial fraction expansion back to the polynomial ratio $B(s)/A(s)$.

The command **printsys (num,den,'s')** prints the num/den in terms of the ratio of polynomials in s.

The command **ilaplace** will find the inverse Laplace transform of a Laplace function.

10.11.1 FINDING ZEROS AND POLES OF B(s)/A(s)

The MATLAB command **[z,p,k] = tf2zp(num,den)** is used to find the zeros, poles, and gain K of $B(s)/A(s)$.

If the zeros, poles, and gain K are given, the following MATLAB command can be used to find the original num/den:

$$[num,den] = zp2tf\ (z,p,k)$$

MODEL PROBLEMS AND SOLUTIONS

Example 2.7. *Consider the function*

$$H(s) = \frac{n(s)}{d(s)}$$

where $n(s) = s^4 + 6s^3 + 5s^2 + 4s + 3$

$d(s) = s^5 + 7s^4 + 6s^3 + 5s^2 + 4s + 7$

(a) *Find n(– 10), n(– 5), n(– 3) and n(– 1)*

(b) *Find d(– 10), d(– 5), d(– 3) and d(– 1)*

(c) *Find H(– 10), H(– 5), H(– 3) and H(–1)*

Solution:

(a)
```
>> n=[1 6 5 4 3];    % n=s^4+6s^3+5s^2+4s+3
>> d=[1 7 6 5 4 7];    % d=s^5+7s^4+6s^3+5s^2+4s+7
>> n2=polyval(n,[-10])
n2 = 4463
>> nn10=polyval(n,[-10])
nn10 = 4463
>> nn5=polyval(n,[-5])
nn5 = -17
>> nn3=polyval(n,[-3])
nn3 = -45
>> nn1=polyval(n,[-1])
nn1 = -1
```

(b)
```
>> dn10=polyval(d,[-10])
dn10 = -35533
>> dn5=polyval(d,[-5])
dn5 = 612
>> dn3=polyval(d,[-3])
```

```
        dn3 = 202
        >> dn1=polyval(d,[-1])
        dn1 = 8
(c)     >> Hn10=nn10/dn10
        Hn10 = -0.1256
        >> Hn5=nn5/dn5
        Hn5 = -0.0278
        >> Hn3=nn3/dn3
        Hn3 = -0.2228
        >> Hn1=nn1/dn1
        Hn1 = -0.1250
```

Example 2.8. *Generate a plot of*

$$y(x) = e^{-0.7x} \sin \omega x$$

where $\omega = 15 \; rad/s$, *and* $0 \leq x \leq 15$. *Use the colon notation to generate the x vector in increments of 0.1.*

Solution:

```
        >> x = [0 : 0.1 : 15];
        >> w = 15;
        >> y = exp(-0.7*x).*sin(w*x);
        >> plot(x,y)
        >> title('y(x) = e^-^0^.^7^xsin \omegax')
        >> xlabel('x')
        >> ylabel('y')
```

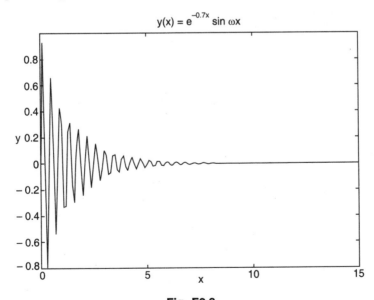

Fig. E2.8

Example 2.9. *Generate a plot of*

$$y(x) = e^{-0.6x} \cos \omega x$$

where $\omega = 10$ *rad/s, and* $0 \le x \le 15$. *Use the colon notation to generate the x vector in increments of 0.05.*

Solution:

```
>> x = [0 : 0.1 : 15];
>> w = 10;
>> y = exp(-0.6*x).*cos(w*x);
>> plot(x,y)
>> title('y(x) = e^-^0^.^6^xcos \omegax')
>> xlabel('x')
>> ylabel('y')
```

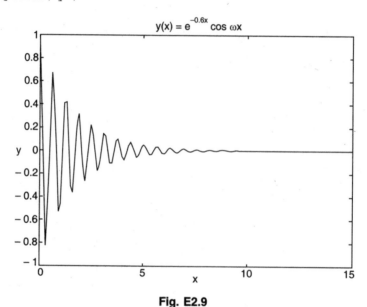

Fig. E2.9

Example 2.10. *Using the functions for plotting x-y data given in Table 2.29 plot the following functions.*

(a) $r^2 = 5 \cos 3t$ $0 \le t \le 2\pi$

(b) $r^2 = 5 \cos 3t$ $0 \le t \le 2\pi$

    $x = r \cos t, \ y = r \sin t$

(c) $y_1 = e^{-2x} \cos x$ $0 \le t \le 20$

    $y_2 = e^{2x}$

(d) $y = \dfrac{\cos(x)}{x}$ $-5 \le x \le 5\pi$

(e) $f = e^{-3t/5} \cos t$ $0 \le t \le 2\pi$

(f) $z = -\dfrac{1}{3}x^2 + 2xy + y^2$

    $|x| \le 7, \ |y| \le 7$

Solution:

(*a*)
```
t = linspace(0, 2*pi, 200);
r =sqrt(abs(5*cos(3*t)));
polar(t,r)
```

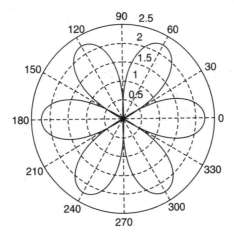

Fig. E2.10(a)

(*b*)
```
t = linspace(0, 2*pi, 200);
r =sqrt(abs(5*cos(3*t)));
x=r.*cos(t);
y=r.*sin(t);
fill(x,y,'k'),
axis('square')
```

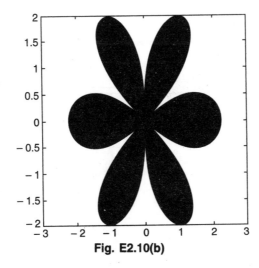

Fig. E2.10(b)

(*c*)
```
x=1:0.1:20;
y1=exp(-2*x).*cos(x);
y2=exp(2*x);
Ax=plotyy(x,y1,x,y2);
hy1=get(Ax(1),'ylabel');
hy2=get(Ax(2),'ylabel');
set(hy1,'string','exp(-2x).cos(x)')
set(hy2,'string','exp(-2x)');
```

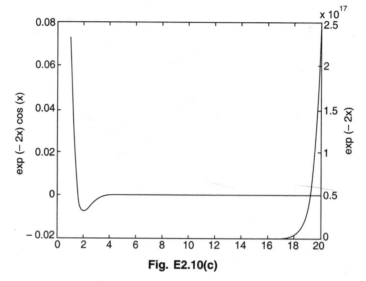

Fig. E2.10(c)

(*d*)
```
x=linspace(-5*pi,5*pi,100);
y=cos(x)./x;
area(x,y);
xlabel('x (rad)'),ylabel('cos(x)/x')
hold on
```

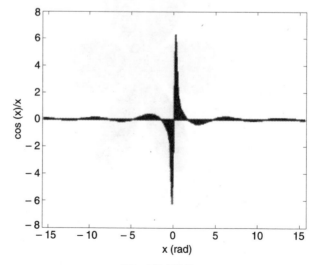

Fig. E2.10(d)

(e) ```
     t=linspace(0,2*pi,200);
     f=exp(-0.6*t).*sin(t);
     stem(t,f)
     ```

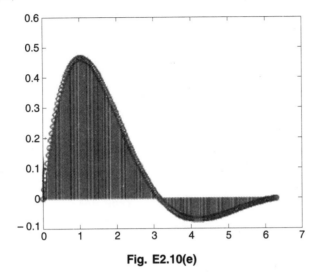

**Fig. E2.10(e)**

(f)  ```
     r=-7:0.2:7;
     [X,Y]=meshgrid(r,r);
     Z=-0.333*X.^2+2*X.*Y+Y.^2;
     cs=contour(X,Y,Z);
     label(cs)
     ```

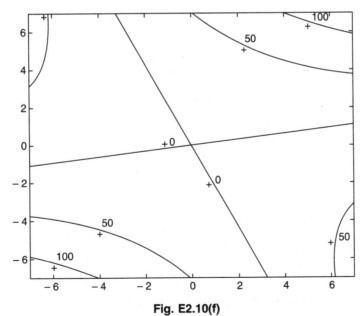

Fig. E2.10(f)

Example 2.11. *Use the functions listed in Table 2.30 for plotting 3-D data for the following.*

(a) $z = \cos x \cos y \ e^{-\frac{\sqrt{x^2 + y^2}}{5}}$

   $|x| \leq = 7, \ |y| \leq 7$

(b) *Discrete data plots with stems*

   $x = t, y = t \cos(t)$

   $z = e^{t/5} - 2 \quad 0 \leq t \leq 5\pi$

(c) *A cylinder generated by*

   $$r = \sin(5\pi z) + 3$$

   $$0 \leq z \leq 1 \quad 0 \leq \theta \leq 2\pi$$

Solution:

(a)
```
u=-7:0.2:7;
[X,Y]=meshgrid(u,u);
Z=cos(X).*cos(Y).*exp(-sqrt(X.^2+Y.^2)/5);
surf(X,Y,Z)
```

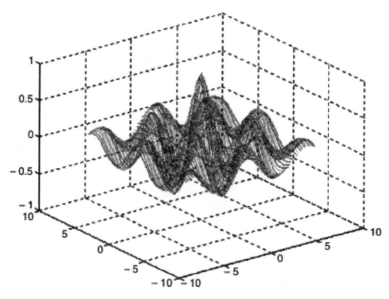

Fig. E2.11(a)

(b)
```
t=linspace(0,5*pi,200);
x=t;y=t.*cos(t);
z=exp(t/5)-2;
stem3(x,y,z,'filled');
xlabel('t'),ylabel('tcos(t)'),zlabel('e^t/5-1')
```

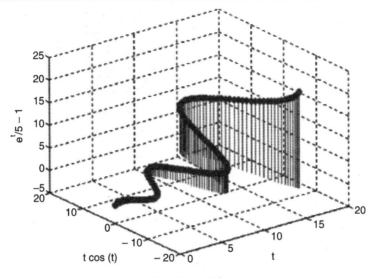

Fig. E2.11(b)

(c)
```
z=[0:0.2:1]';
r=sin(5*pi*z)+3;
cylinder(r)
```

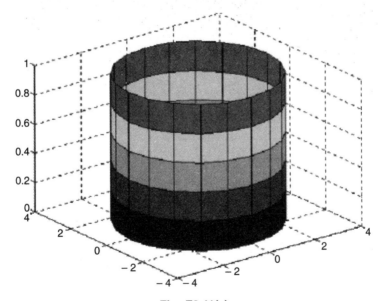

Fig. E2.11(c)

Example 2.12. *Obtain the plot of the points for $0 \leq t \leq 6\pi$ when the coordinates x,y,z are given as a function of the parameter t as follows:*

$$x = \sqrt{t}\ sin\ (3t)$$

$$y = \sqrt{t}\ cos\ (3t)$$

$$z = 0.8t$$

Solution:

% Line plots

```
>> t=[0:0.1:6*pi];
>> x=sqrt(t).*sin(3*t);
>> y=sqrt(t).*cos(3*t);
>> z=0.8*t;
>> plot3(x,y,z,'k','linewidth',1)
>> grid on
>> xlabel('x');ylabel('y');zlabel('z')
```

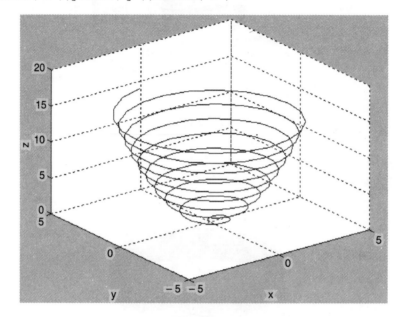

Fig. E2.12

Example 2.13. *Obtain the mesh and surface plots for the function* $z = \dfrac{2xy^2}{x^2 + y^2}$ *over the domain* $-2 \le x \le 6$ *and* $2 \le y \le 8$.

Solution:

% Mesh and surface plots

```
x=-2:0.1:6;
>> y=2:0.1:8;
>> [x,y]=meshgrid(x,y);
>> z=2*x.*y.^2./(x.^2+y.^2);
>> mesh(x,y,z)
>> xlabel('x');ylabel('y');zlabel('z')
>> surf(x,y,z)
>> xlabel('x');ylabel('y');zlabel('z')
```

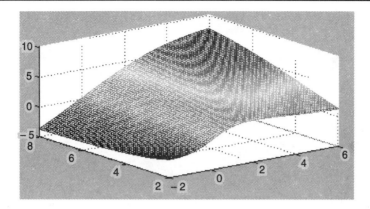

Fig. E2.13 (a)

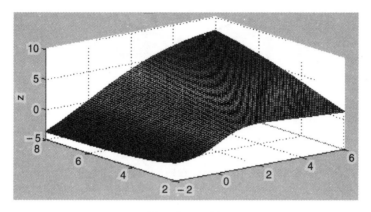

Fig. E2.13 (b)

Example 2.14. *Plot the function* $z = 2^{-1.5\sqrt{x^2 + y^2}} \sin(x) \cos(0.5\,y)$ *over the domain* $-4 \leq x \leq 4$ *and* $-4 \leq y \leq 4$ *using Table 2.30.*

(a) *Mesh plot*

(b) *Surface plot*

(c) *Mesh curtain plot*

(d) *Mesh and contour plot*

(e) *Surface and contour plot*

Solution:

(a)   % Mesh Plot

```
>> x=-4:0.25:4;
>> y=-4:0.25:4;
>> [x,y]=meshgrid(x,y);
>> z=2.^(-1.5*sqrt(x.^2 + y.^2)).*cos(0.5*y).*sin(x);
>> mesh(x,y,z)
>> xlabel('x');ylabel('y')
>> zlabel('z')
```

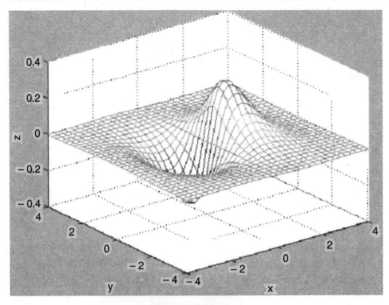

Fig. E2.14 (a)

(*b*)   % Surface Plot

```
>> x=-4:0.25:4;
>> y=-4:0.25:4;
>> [x,y]=meshgrid(x,y);
>> z=2.0.^(-1.5*sqrt(x.^2+y.^2)).*cos(0.5*y).*sin(x);
>> surf(x,y,z)
>> xlabel('x');ylabel('y')
>> zlabel('z')
```

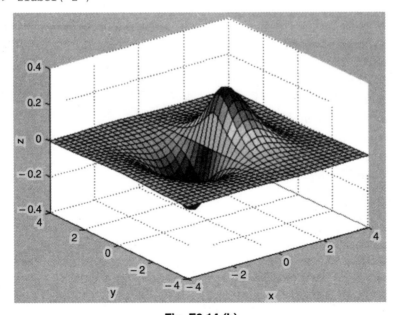

Fig. E2.14 (b)

(*c*)   % Mesh Curtain Plot

```
>> x=-4.0:0.25:4;
>> y=-4.0:0.25:4;
>> [x,y]=meshgrid(x,y);
>> z=2.0.^(-1.5*sqrt(x.^2+y.^2)).*cos(0.5*y).*sin(x);
>> meshz(x,y,z)
>> xlabel('x');ylabel('y')
>> zlabel('z')
```

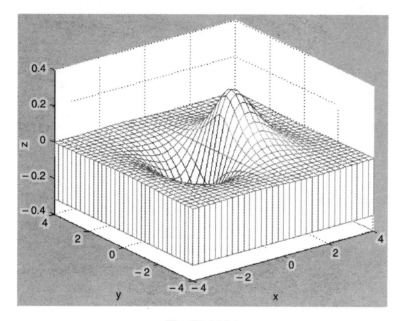

Fig. E2.14 (c)

(*d*)   % Mesh and Contour Plot

```
>> x=-4.0:0.25:4;
>> y=-4.0:0.25:4;
>> [x,y]=meshgrid(x,y);
>> z=2.0.^(-1.5*sqrt(x.^2+y.^2)).*cos(0.5*y).*sin(x);
>> meshc(x,y,z)
>> xlabel('x');ylabel('y')
>> zlabel('z')
```

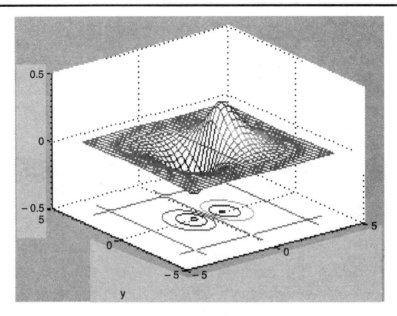

Fig. E2.14 (d)

(*e*)   % Surface and Contour Plot

```
>> x=-4.0:0.25:4;
>> y=-4.0:0.25:4;
>> [x, y] =meshgrid(x, y);
>> z=2.0. ^ (-1.5*sqrt (x. ^2+y. ^2)).*cos (0.5*y).*sin(x);
>> surfc(x, y, z)
>> xlabel('x');ylabel('y')
>> zlabel('z')
```

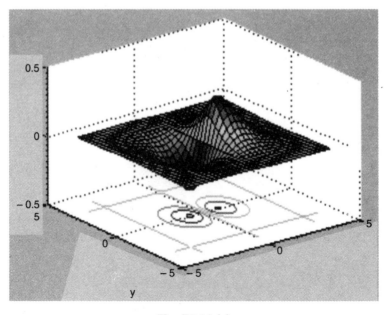

Fig. E2.14 (e)

Example 2.15. *Plot the function* $z = 2^{-1.5\sqrt{x^2 + y^2}}$ *sin* (x) *cos* $(0.5\,y)$ *over the domain* $-4 \le x$ ≤ 4 *and* $-4 \le y \le 4$ *using Table 2.30.*

(*a*) *Surface plot with lighting*

(*b*) *Waterfall plot*

(*c*) *3-D contour plot*

(*d*) *2-D contour plot*

Solution:

(*a*)   % Surface Plot with Lighting

```
>> x=-4.0:0.25:4;
>> y=-4.0:0.25:4;
>> [x,y]=meshgrid(x,y);
>> z=2.0.^(-1.5*sqrt(x.^2+y.^2)).*cos(0.5*y).*sin(x);
>> surfl(x,y,z)
>> xlabel('x');ylabel('y')
>> zlabel('z')
```

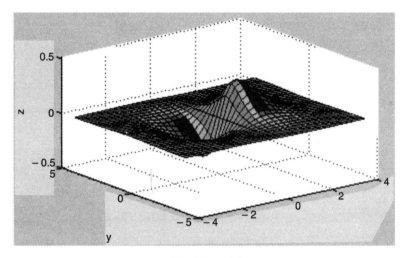

Fig. E2.15 (a)

(*b*)   % Waterfall Plot

```
>> x=-4.0:0.25:4;
>> y=-4.0:0.25:4;
>> [x,y]=meshgrid(x,y);
>> z=2.0.^(-1.5*sqrt(x.^2+y.^2)).*cos(0.5*y).*sin(x);
>> waterfall(x,y,z)
>> xlabel('x');ylabel('y')
>> zlabel('z')
```

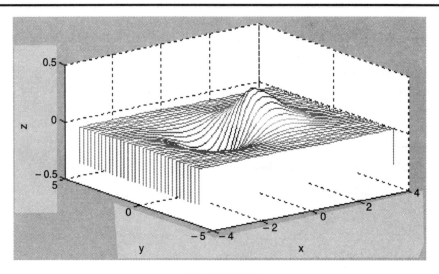

Fig. E2.15 (b)

(*c*)   % 3-D Contour Plot

```
>> x=-4.0:0.25:4;
>> y=-4.0:0.25:4;
>> [x,y]=meshgrid(x,y);
>> z=2.0.^(-1.5*sqrt(x.^2+y.^2)).*cos(0.5*y).*sin(x);
>> contour3(x,y,z,15)
>> xlabel('x');ylabel('y')
>> zlabel('z')
```

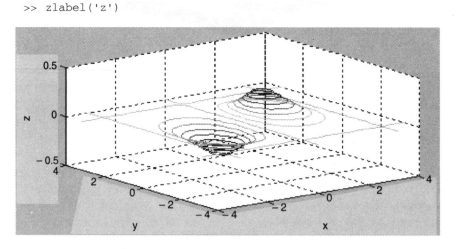

Fig. E2.15 (c)

(*d*)   % 2-D Contour Plot

```
>> x=-4.0:0.25:4;
>> y=-4.0:0.25:4;
>> [x,y]=meshgrid(x,y);
>> z=2.0.^(-1.5*sqrt(x.^2+y.^2)).*cos(0.5*y).*sin(x);
>> contour(x,y,z,15)
```

```
>> xlabel('x');ylabel('y')
>> zlabel('z')
```

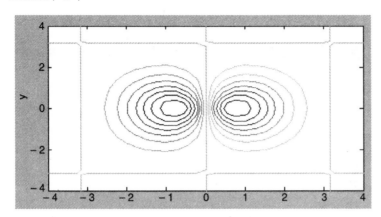

Fig. E2.15 (d)

Example 2.16. *Using the functions given in Table 2.29 for plotting x-y data, plot the following functions:*

(a) $f(t) = t \cos t$ $0 \le t \le 10\pi$

(b) $x = e^{-2t}, y = t$ $0 \le t \le 2\pi$

(c) $x = t, y = e^{2t}$ $0 \le t \le 2\pi$

(d) $x = e^t, y = 50 + e^t$ $0 \le t \le 2\pi$

(e) $\begin{aligned} r^2 &= 3 \sin 7t \\ y &= r \sin t \end{aligned}$ $0 \le t \le 2\pi$

(f) $\begin{aligned} r^2 &= 3 \sin 4t \\ y &= r \sin t \end{aligned}$ $0 \le t \le 2\pi$

(g) $y = t \sin t$ $0 \le t \le 5\pi$

Solution:

(a)   % Use of Plot Command

```
>> fplot('x.*cos(x)',[0,10*pi])
```

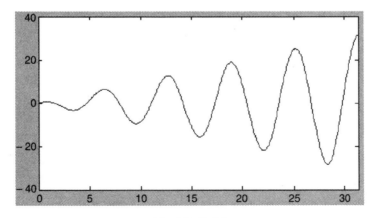

Fig. E2.16 (a)

(b)   % Semilog x Command

```
>> t=linspace(0,2*pi,200);
>> x=exp(-2*t);y=t;
>> semilogx(x,y),grid
```

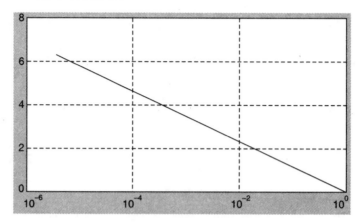

Fig. E2.16 (b)

(c)   % Semilog y Command

```
t=linspace(0,2*pi,200);
>> semilogy(t,exp(-2*t)),grid
```

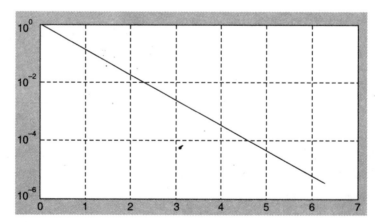

Fig. E2.16 (c)

(d)   % Use of loglog Command

```
>> t=linspace(0,2*pi,200);
>> x=exp(t);
>> y=50+exp(t);
>> loglog(x,y),grid
```

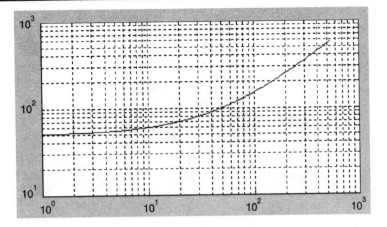

Fig. E2.16 (d)

(*e*) %Use of stairs Command

```
>> t=linspace(0,2*pi,200);
>> r=sqrt(abs(3*sin(7*t)));
>> y=r.*sin(t);
>> stairs(t,y)
>> axis([0 pi 0 inf]);
```

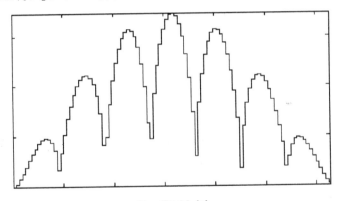

Fig. E2.16 (e)

(*f*) % Use of bar Command

```
>> t=linspace(0,2*pi,200);
>> r=sqrt(abs(3*sin(4*t)));
>> y=r.*sin(t);
>> bar(t,y)
>> axis([0 pi 0 inf]);
```

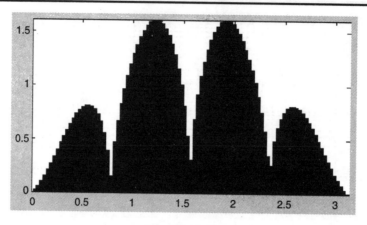

Fig. E2.16 (f)

(g) %use of comet Command

```
>> q=linspace(0,5*pi,200);
>> y=q.*sin(q);
>> comet(q,y)
```

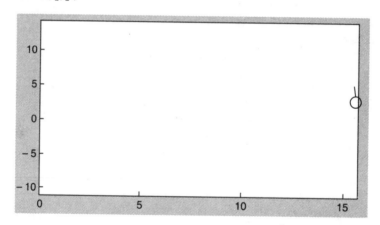

Fig. E2.16 (g)

Example 2.17. *Consider the two matrices*

$$A = \begin{bmatrix} 3 & 2\pi \\ 5j & 10 + \sqrt{2}j \end{bmatrix}$$

$$B = \begin{bmatrix} 7j & -15j \\ 2\pi & 18 \end{bmatrix}$$

Using MATLAB, determine the following:

(a) $A + B$

(b) AB

(c) A^2

(d) A^T

(e) B^{-1}

(f) $B^T A^T$

(g) $A^2 + B^2 - AB$

Solution:

```
>> A = [3 2*pi;5j 10+sqrt(2)*j];
>> B = [7j -15j;2*pi 18];
```

(a) A + B

```
ans =

     3.0000 + 7.0000i    6.2832 -15.0000i
     6.2832 + 5.0000i   28.0000 + 1.4142i
```

(b) >> A * B

```
ans =

   1.0e+002 *

     0.3948 + 0.2100i    1.1310 - 0.4500i
     0.2783 + 0.0889i    2.5500 + 0.2546i
```

(c) >> A^2

```
ans =

     9.0000  +31.4159i   81.6814 + 8.8858i
    -7.0711  +65.0000i   98.0000 +59.7002i
```

(d) >> inv(A)

```
ans =

     0.1597 + 0.1917i   -0.1150 - 0.1042i
     0.0829 - 0.0916i    0.0549 + 0.0498i
```

(e) >> B^-1

```
ans =

     0 - 0.0817i    0.0681
     0 + 0.0285i    0.0318
```

(f) >> inv(B) * inv(A)

```
ans =

     0.0213 - 0.0193i   -0.0048 + 0.0128i
    -0.0028 + 0.0016i    0.0047 - 0.0017i
```

(g)   >> (A^2 + B^2) - (A * B)

```
ans =

   1.0e+002*

    -0.7948 - 0.8383i    0.7358 - 2.1611i
     0.7819 + 1.0010i    1.6700 - 0.6000i
```

Example 2.18. *Find the inverse of the following matrices using MATLAB:*

(a) $\begin{bmatrix} 3 & 2 & 0 \\ 2 & -1 & 7 \\ 5 & 4 & 9 \end{bmatrix}$ (b) $\begin{bmatrix} -4 & 2 & 5 \\ 7 & -1 & 6 \\ 2 & 3 & 7 \end{bmatrix}$

(c) $\begin{bmatrix} -1 & 2 & -5 \\ 4 & 3 & 7 \\ 7 & -6 & 1 \end{bmatrix}$

Solution:

```
>> clear % Clears the workspace
>> A = [3 2 0; 2 -1 7; 5 4 9]; % Spaces separate matrix columns - semicolons
separate matrix rows
>> B = [-4 2 5; 7 -1 6; 2 3 7]; % Spaces separate matrix columns - semicolons
separate matrix rows
>> C = [-1 2 -5; 4 3 7; 7 -6 1]; % Spaces separate matrix columns -
semicolons separate matrix rows
>> inv(A); % Finds the inverse of the selected matrix
>> inv(B); % Finds the inverse of the selected matrix
>> inv(C) % Finds the inverse of the selected matrix
% Inverse of A
ans =
      0.4805      0.2338     -0.1818
     -0.2208     -0.3506      0.2727
     -0.1688      0.0260      0.0909
% Inverse of B
ans =
     -0.1773      0.0071      0.1206
     -0.2624     -0.2695      0.4184
      0.1631      0.1135     -0.0709
% Inverse of  C
ans =
      0.1667      0.1037      0.1074
      0.1667      0.1259     -0.0481
     -0.1667      0.0296     -0.0407
```

Example 2.19. *Determine the eigenvalues and eigenvectors of matrix* ***A*** *using MATLAB.*

$$(a)\, A = \begin{bmatrix} 4 & -1 & 5 \\ 2 & 1 & 3 \\ 6 & -7 & 9 \end{bmatrix}$$

$$(b)\, A = \begin{bmatrix} 3 & 5 & 7 \\ 2 & 4 & 8 \\ 5 & 6 & 10 \end{bmatrix}$$

Solution:

```
(a) A = [4 - 1 5 ; 2 1 3 ; 6 - 7 9]
A =
      4      -1      5
      2       1      3
      6      -7      9
%The eigenvalues of A
format short e
eig(A)
ans =
```

```
    1.0000e+001
    5.8579e-001
    3.4142e+000
%The eigenvectors of A
[Q,d]=eig(A)
Q =
 -5.5709e-001  -8.2886e-001  -7.3925e-001
 -3.7139e-001  -3.9659e-002  -6.7174e-001
 -7.4278e-001   5.5805e-001  -4.7739e-002
d =
    1.0000e+001   0            0
        0     5.8579e-001      0
        0         0        3.4142e+000
```

(b) A =

```
        3       5       7
        2       4       8
        5       6      10
%The eigenvalues of A
format short e
eig(A)
ans =
    1.7686e+001
   -3.4295e-001 +1.0066e+000i
   -3.4295e-001 -1.0066e+000i
%The eigenvectors of A
[Q,d]=eig(A)
Q =
Column 1
    5.0537e-001
    4.8932e-001
    7.1075e-001
Column 2
   -2.0715e-001 -5.2772e-001i
    7.1769e-001
   -3.3783e-001 +2.2223e-001i
Column 3
   -2.0715e-001 +5.2772e-001i
    7.1769e-001
   -3.3783e-001 -2.2223e-001i
d =
Column 1
1.7686e+001
          0
          0
```

```
Column 2
    0
 -3.4295e-001  +1.0066e+000i
    0


Column 3
    0
    0
 -3.4295e-001  -1.0066e+000i
```

Example 2.20. *Determine the eigenvalues and eigenvectors of* **AB** *using MATLAB.*

$$A = \begin{bmatrix} 3 & 0 & 2 & 1 \\ 1 & 2 & 5 & 4 \\ 7 & -1 & 2 & 6 \\ 1 & -2 & 3 & 4 \end{bmatrix}$$

$$B = \begin{bmatrix} 1 & 3 & 5 & 7 \\ 2 & -1 & -2 & 4 \\ 3 & 2 & 1 & 1 \\ 4 & 1 & 0 & 6 \end{bmatrix}$$

Solution:

% MATLAB Program

*% The matrix "a" = A*B*

```
>> A = [3 0 2 1; 1 2 5 4; 7 - 1 2 6; 1 - 2 3 4];
>> B = [1 3 5 7; 2 - 1 - 2 4; 3 2 1 1; 4 1 0 6];
>> a = A*B
a =
      13    14    17    29
      36    15     6    44
      35    32    39    83
      22    15    12    26
>> eig (a)
Ans. =
98.5461
2.2964
-1.3095
-6.5329
```

The eigenvectors are :

```
>> [Q, d] = eig (a)
Q =
      -0.3263    -0.2845     0.3908     0.3413
      -0.3619     0.7387    -0.7816    -0.9215
      -0.8168    -0.6026     0.4769     0.0962
      -0.3089     0.1016    -0.0950     0.1586
```

d =

98.5461	0	0	0
0	2.2964	0	0
0	0	-1.3095	0
0	0	0	-6.5329

Example 2.21. *Solve the following set of equations using MATLAB.*

(a) $x_1 + 2x_2 + 3x_3 + 5x_4 = 21$

$-2x_1 + 5x_2 + 7x_3 - 9x_4 = 18$

$5x_1 + 7x_2 + 2x_3 - 5x_4 = 25$

$-x_1 + 3x_2 - 7x_3 + 7x_4 = 30$

(b) $x_1 + 2x_2 + 3x_3 + 4x_4 = 8$

$2x_1 - 2x_2 - x_3 - x_4 = -3$

$x_1 - 3x_2 + 4x_3 - 4x_4 = 8$

$2x_1 + 2x_2 - 3x_3 + 4x_4 = -2$

Solution:

(a)

```
>> A= [1 2 3 5;-2 5 7 -9; 5 7 2 -5;-1 -3 -7 7];
>> B = [21; 18; 25; 30];
>> S = A\B
S =
 - 8.9896
  14.1285
 - 5.4438
  3.6128
```

% Therefore $x_1 = -8.9896$, $x_2 = 14.12.85$, $x_3 = -5.4438$, $x_4 = 3.6128$.

(b)

```
>> A= [1 2 3 4; 2 -2 -1 1; 1 -3 4 -4; 2 2 -3 4];
>> B = [8;-3; 8;-2];
>> S=A\B
S =
  2.0000
  2.0000
  2.0000
 - 1.0000
```

%Therefore $x_1 = 2.0000$, $x_2 = 2.0000$, $x_3 = 2.0000$, $x_4 = -1.0000$.

Example 2.22. *Use diff command for symbolic differentiation of the following functions:*

(a) $S_1 = e^{x^8}$

(b) $S_2 = 3x^3 e^{x^5}$

(c) $S_3 = 5x^3 - 7x^2 + 3x + 6$

Solution:

(a)

```
>> syms x
>> S1=exp(x^8);
>> diff (S1)
ans =
8*x^7*exp(x^8)
```

(b)

```
>> S2=3*x^3*exp(x^5);
>> diff (S2)
ans =
9*x^2*exp(x^5)+15*x^7*exp(x^5)
```

(c)

```
>> S3=5*x^3-7*x^2+3*x+6;
>> diff (S3)
ans =
15*x^2-14*x+3
```

Example 2.23. *Use MATLAB's symbolic commands to find the values of the following integrals.*

(a) $\int_{0.2}^{0.7} |x| \, dx$

(b) $\int_{0}^{\pi} (\cos y + 7y^2) \, dy$

(c) $\sqrt{x}$

(d) $7x^5 - 6x^4 + 11x^3 + 4x^2 + 8x + 9$

(e) $\cos a$

Solution:

(a)

```
>>syms x, y, a, b
>> S1=abs(x)
>> int (S1, 0.2, 0.7)
ans =
9/40
```

(b)

```
>> S2=cos (y) +7*y^2
>> int (S2, 0, pi)
ans =
7/3*pi^3
```

(c)

```
>> S3=sqrt (x)
>> int (S3)
ans =
2/3*x^ (3/2)
```

```
      >> int (S3,'a','b')
 ans =
 2/3*b^ (3/2)-2/3*a^ (3/2)
 >> int (S3, 0.4, 0.7)
 ans =
 7/150*70^ (1/2)-4/75*10^ (1/2)
```
(*d*)
```
      >> S4=7*x^5-6*x^4+11*x^3+4*x^2+8*x-9
      >> int (S4)
 ans =
 7/6*x^6-6/5*x^5+11/4*x^4+4/3*x^3+4*x^2-9*x
```
(*e*)
```
      >> S5=cos (a)
      >> int (S5)
 ans =
 sin (a)
```

Example 2.24. *Obtain the general solution of the following first order differential equations:*

(*a*) $\dfrac{dy}{dt} = 5t - 6y$

(*b*) $\dfrac{d^2y}{dt^2} + 3\dfrac{dy}{dt} + y = 0$

(*c*) $\dfrac{ds}{dt} = Ax^3$

(*d*) $\dfrac{ds}{dA} = Ax^3$

Solution:

(*a*)
```
      >> solve ('Dy=5*t-6*y')
 ans =
 5/6*t-5/36+exp (-6*t)*C1
```
(*b*)
```
      >> dsolve ('D2y+3*Dy+y=0')
 ans =
 C1*exp (1/2*(5^ (1/2)-3)*t) +C2*exp (-1/2*(5^ (1/2) +3)*t)
```
(*c*)
```
      >> dsolve ('Ds=A*x^3','x')
 ans =
 1/4*A*x^4+C1
```
(*d*)
```
      >> dsolve ('Ds=A*x^3','A')
 ans =
 1/2*A^2*x^3+C1
```

Example 2.25. *Determine the solution of the following differential equations that satisfies the given initial conditions.*

(a) $\dfrac{dy}{dx} = -7x^2$ $y(1) = 0.7$

(b) $\dfrac{dy}{dx} = 5x \cos^2 y$ $y(0) = \pi/4$

(c) $\dfrac{dy}{dx} = -y + e^{3x}$ $y(0) = 2$

(d) $\dfrac{dy}{dt} + 5y = 35$ $y(0) = 4$

Solution:

(a)
```
>> dsolve ('Dy=-7*x^2','y (1) =0.7')
ans =
-7*x^2*t+7*x^2+7/10
```
(b)
```
>> dsolve ('Dy=5*x*cos (y) ^2','y (0) =pi/4')
ans =
atan (5*t*x+1)
```
(c)
```
>> dsolve ('Dy=-y+ exp (3*x)','y (0) =2')
ans =
exp (3*x) +exp (-t)*(-exp (3*x) +2)
```
(d)
```
>> dsolve ('Dy+5*y=35','y (0) =4')
ans =
7-3*exp (-5*t)
```

Example 2.26. *Given the differential equation*

$$\frac{d^2x}{dt^2} + 7\frac{dx}{dt} + 5x = 8u(t) \qquad t \ge 0$$

Using MATLAB program, find

(a) x(t) when all the initial conditions are zero

(b) x(t) when x(0) = 1 and $\dot{x}(0) = 2$.

Solution:

(a) x(t) when all the initial conditions are zero
```
>> x = dsolve ('D2x = -7*Dx - 5*x +8', 'x (0) = 0')
x =
8/5+ (-8/5-C2)*exp (1/2*(-7+29^ (1/2))*t) +C2*exp (-1/2*(7+29^ (1/2))*t)
```
(b) x(t) when x(0) = 1 and $\dot{x}(0) = 2$
```
>> x = dsolve ('D2x = -7*Dx - 5*x +8', 'x (0) = 1', 'Dx (0) = 2')
x =
```

8/5+ (-3/10-1/290*29^ (1/2))*exp (1/2*(-7+29^ (1/2))*t)-1/290*(-1+3*29^
(1/2))*29^ (1/2)*exp (-1/2*(7+29^ (1/2))*t)

Example 2.27. *Given the differential equation*

$$\frac{d^2x}{dt^2} + 12\,\frac{dx}{dt} + 15x = 35 \qquad t \geq 0$$

Using MATLAB program, find

(a) x(t) when all the initial conditions are zero

(b) x(t) when x(0) = 0 and ẋ(0) = 1.

Solution:

(a) x(t) when all the initial conditions are zero

```
>> x = dsolve ('D2x = -12*Dx - 15*x +35', 'x (0) = 0')
x =
7/3+ (-7/3-C2)*exp ((-6+21^ (1/2))*t) +C2*exp (-(6+21^ (1/2))*t)
```

(b) x(t) when x(0) = 0 and ẋ(0) = 1

```
>> x = dsolve ('D2x = -12*Dx - 15*x + 35', 'x (0) = 0', 'Dx (0) = 1')
x =
7/3+ (-7/6-13/42*21^ (1/2))*exp ((-6+21^ (1/2))*t)-1/126*
(-39+7*21^ (1/2))*21^ (1/2)*exp (-(6+21^ (1/2))*t)
```

Example 2.28. *Find the inverse of the following matrix using MATLAB.*

$$A = \begin{bmatrix} s & 2 & 0 \\ 2 & s & -3 \\ 3 & 0 & 1 \end{bmatrix}$$

Solution:

```
>> A = [s 2 0; 2 s -3; 3 0 1];
>> inv (A)
ans =
    [s/(s^2-22),    -2/(s^2-22),    -6/(s^2-22)]
    [-11/(s^2-22),  s/(s^2-22),     3*s/(s^2-22)]
    [-3*s/(s^2-22), 6/(s^2-22),     (s^2-4)/(s^2-22)]
```

Example 2.29. *Expand the following function F(s) into partial fractions using MATLAB. Determine the inverse Laplace transform of F(s).*

$$F(s) = \frac{1}{s^4 + 5s^3 + 7s^2}$$

The MATLAB program for determining the partial-fraction expansion is given below:

Solution:

```
>> b = [0 0 0 0 1];
>> a = [1 5 7 0 0];
>> [r, p, k] = residue (b, a)
r =
        0.0510 - 0.0648i
        0.0510 + 0.0648i
```

```
            -0.1020
             0.1429
    p =
          -2.5000 + 0.8660i
          -2.5000 - 0.8660i
               0
               0
  k = [ ]
```

% From the above MATLAB output, we have the following expression:

$$F(s) = \frac{r_1}{s - p_1} + \frac{r_2}{s - p_2} + \frac{r_3}{s - p_3} + \frac{r_4}{s - p_4} =$$

$$F(s) = \frac{0.0510 - 0.0648i}{s - (-2.5000 + 0.8660i)} + \frac{(0.0510 + 0.0648i}{s - (-2.5000 - 0.8660i)} + \frac{-0.1020}{s - 0} + \frac{0.1429}{s - 0}$$

% Note that the row vector k is zero implies that there is no constant term in this example problem.

% The MATLAB program for determining the inverse Laplace transform of $F(s)$ is given below:

```
>> syms s
>> f = 1/(s^4 + 5*s^3 + 7*s^2);
>> ilaplace (f)
ans =
1/7*t-5/49+5/49*exp (-)*cos (1/2*3^ (1/2)*t) +11/147*exp (-5/2*t)*3^
(1/2)*sin(1/2*3^(1/2)*t)
```

Example 2.30. *Expand the following function F(s) into partial fractions using MATLAB. Determine the inverse Laplace transform of F(s).*

$$F(s) = \frac{5s^2 + 3s + 6}{s^4 + 3s^3 + 7s^2 + 9s + 12}$$

Solution:

The MATLAB program for determining the partial-fraction expansion is given below:

```
>> b = [0 0 5 3 6];
>> a = [1 3 7 9 12];
>> [r,p,k] = residue(b,a)
r =
          -0.5357 - 1.0394i
          -0.5357 + 1.0394i
           0.5357 - 0.1856i
           0.5357 + 0.1856i
    p =
          -1.5000 + 1.3229i
          -1.5000 - 1.3229i
          -0.0000 + 1.7321i
          -0.0000 - 1.7321i
  k = [ ]
```

% From the above MATLAB output, we have the following expression:

$$F(s) = \frac{r_1}{s - p_1} + \frac{r_2}{s - p_2} + \frac{r_3}{s - p_3} + \frac{r_4}{s - p_4} =$$

$$F(s) = \frac{-0.5357 - 0.0394i}{s - (-1.500 + 1.3229i)} + \frac{(-0.5357 + 1.0394i)}{s - (-1.500 + 1.3229i)}$$

$$+ \frac{0.5357 - 0.1856i}{s - (-0 + 1.7321i)} + \frac{0.5357 - 0.1856i}{s - (-0 - 1.7321i)}$$

% Note that the row vector k is zero implies that there is no constant term in this example problem.

% The MATLAB program for determining the inverse Laplace transform of $F(s)$ is given below:

```
>> syms s
>> f = (5*s^2 + 3*s +6)/(s^4 + 3*s^3 + 7*s^2 + 9*s +12);
>> ilaplace(f)
ans =
11/14*exp(-3/2*t)*7^(1/2)*sin(1/2*7^(1/2)*t)-15/14*exp(-3/2*t)*cos
(1/2*7^(1/2)*t)+3/14*3^(1/2)*sin(3^(1/2)*t)+15/14*cos(3^(1/2)*t)
```

Example 2.31. *For the following function F(s):*

$$F(s) = \frac{s^4 + 3s^3 + 5s^2 + 7s + 25}{s^4 + 5s^3 + 20s^2 + 40s + 45}$$

Using MATLAB, find the partial-fraction expansion of F(s). Also, find the inverse Laplace transformation of F(s).

Solution:

$$F(s) = \frac{s^4 + 3s^3 + 5s^2 + 7s + 25}{s^4 + 5s^3 + 20s^2 + 40s + 45}$$

The partial-fraction expansion of $F(s)$ using MATLAB program is given as follows:

```
num = [ 1   3   5   7   25];
den = [1   5   20   40   45];
[r,p,k]  = residue(num,den)
r =
-1.3849 + 1.2313i
-1.3849 - 1.2313i
 0.3849 - 0.4702i
 0.3849 + 0.4702i
p =
-0.8554 + 3.0054i
-0.8554 - 3.0054i
-1.6446 + 1.3799i
-1.6446 - 1.3799i
k =
    1
```

From the MATLAB output, the partial-fraction expansion of $F(s)$ can be written as follows:

$$F(s) = \frac{r_1}{(s - p_1)} + \frac{r_2}{(s - p_2)} + \frac{r_3}{(s - p_3)} + \frac{r_4}{(s - p_4)} + k$$

$$F(s) = \frac{(-1.3849 + j1.2313)}{(s + 0.8554 - j3.005)} + \frac{(-1.3849 - j1.2313)}{(s + 0.8554 + j3.005)}$$

$$+ \frac{(0.3849 - j0.4702)}{(s + 1.6446 - j1.3799)} + \frac{(0.3849 + j0.4702)}{(s + 1.6446 + j1.3779)} + 1$$

Example 2.32. *Obtain the partial-fraction expansion of the following function using MATLAB:*

$$F(s) = \frac{8(s + 1)(s + 3)}{(s + 2)(s + 4)(s + 6)^2}$$

Solution:

$$F(s) = \frac{8(s + 1)(s + 3)}{(s + 2)(s + 4)(s + 6)^2} = \frac{(8s + 8)(s + 3)}{(s^2 + 6s + 8)(s^2 + 12s + 36)}$$

The partial fraction expansion of $F(s)$ using **MATLAB** program is given as follows:

```
EDU>> num=conv([8 8],[1 3]);
EDU>> den=conv([1 6 8],[1 12 36]);
EDU>> [r,p,k]=residue(num,den)

r =

    3.2500
   15.0000
   -3.0000
   -0.2500

p =

   -6.0000
   -6.0000
   -4.0000
   -2.0000

k = [ ]
```

From the above **MATLAB** result, we have the following expansion:

$$F(s) = \frac{r_1}{(s - p_1)} + \frac{r_2}{(s - p_2)} + \frac{r_3}{(s - p_3)} + \frac{r_4}{(s - p_4)} + k$$

$$F(s) = \frac{3.25}{(s + 6)} + \frac{15}{(s - 15)} + \frac{-3}{(s + 3)} + \frac{-0.25}{(s + 0.25)} + 0$$

It should be noted here that the row vector k is zero, because the degree of the numerator is lower than that of the denominator.

$$F(s) = 3.25e^{-6t} + 15e^{15t} - 3e^{-3t} - 0.25e^{-0.25t}$$

Example 2.33. *Find the Laplace transform of the following function using MATLAB.*

(a) $f(t) = 7t^3 \cos (5t + 60°)$

(b) $f(t) = -7t\, e^{-5t}$

(c) $f(t) = -3 \cos 5t$

(d) $f(t) = t \sin 7t$

(e) $f(t) = 5 e^{-2t} \cos 5t$

(f) $f(t) = 3 \sin (5t + 45^{\circ})$

(g) $f(t) = 5 e^{-3t} \cos (t - 45^{\circ})$

Solution:

% MATLAB Program

```
>> syms t    %tell MATLAB that "t" is a symbol.
>> f = 7 * t^3*cos(5*t + (pi/3)); % define the function.
>> laplace(f)
ans =
-84/(s^2+25)^3*s^2+21/(s^2+25)^2+336*(1/2*s-5/2*3^(1/2))/(s^2+25)
^4*s^3-168*(1/2*s-5/2*3^(1/2))/(s^2+25)^3*s
```

>> pretty(laplace(f)) % the pretty function prints symbolic output

% in a format that resembles typeset mathematics.

$$-84\ \frac{s^2}{(s^2+25)^3} + 21\ \frac{1}{(s^2+25)^2} + 336\ \frac{(1/2\ s - 5/2\ 3^{1/2})\ s}{(s^2+25)^4}$$

$$-168\ \frac{(1/2\ s - 5/2\ 3^{1/2})\ s}{(s^2+25)^3}$$

(b)
```
>>syms t x
>>f = -7*t*exp(-5*t);
>> laplace(f,x)
ans =
-7/(x+5)^2
```

(c)
```
>>syms t x
>>f = -3*cos(5*t);
>> laplace(f,x)
ans =
-3*x/(x^2+25)
```

(d)
```
>>syms t x
>>f = t*sin(7*t);
>> laplace(f,x)
ans =
1/(x^2+49)*sin(2*atan(7/x))
```

(e)
```
>>syms t x
>>f = 5*exp(-2*t)*cos(5*t);
>> laplace(f,x)
```

```
        ans =
        5*(x+2)/((x+2)^2+25)
```
(f)   `>>syms t x`
```
        >>f = 3*sin(5*t + (pi/4));
        >> laplace(f,x)
        ans =
        3*(1/2*x*2^(1/2)+5/2*2^(1/2))/(x^2+25)
```
(g)   `>>syms t x`
```
        >>f = 5*exp(-3*t)*cos(t-(pi/4));
        >> laplace(f,x)
        ans =
            5*(1/2*(x+3)*2^(1/2)+1/2*2^(1/2))/((x+3)^2+1)
```

Example 2.34. *Generate partial-fraction expansion of the following function.*

$$F(s) = \frac{10^5\,(s+7)\,(s+13)}{s(s+25)\,(s+55)\,(s^2+7s+75)\,(s^2+7s+45)}$$

Solution:

Generate the partial fraction expansion of the following function:

```
numg=poly[-7 -13];
numg=poly([-7 -13]);
deng=poly([0 -25 -55 roots([1 7 75])' roots([1 7 45])']);
[numg,deng]=zp2tf(numg',deng',1e5);
Gtf=(numg,deng);
Gtf=tf(numg,deng);
G=zpk(Gtf);
[r,p,k]=residue(numg,deng)
r =
  1.0e-017 *
    0.0000
   -0.0014
    0.0254
   -0.1871
    0.1621
   -0.0001
    0.0000
    0.0011
p =
  1.0e+006 *
    4.6406
    1.4250
    0.3029
    0.0336
    0.0027
    0.0001
```

```
        0.0000
             0
 k = [ ]
```

Example 2.35. *Determine the inverse Laplace transform of the following functions using* MATLAB.

(a) $F(s) = \dfrac{s}{s(s+2)(s+6)}$

(b) $F(s) = \dfrac{1}{s^2(s+5)}$

(c) $F(s) = \dfrac{3s+1}{(s^2+2s+9)}$

(d) $F(s) = \dfrac{s-25}{s(s^2+3s+20)}$

Solution:

(a)
```
>> syms s
>> f = s/(s*((s + 2)*(s + 6)));
>> ilaplace(f)
ans =
1/2*exp(-4*t)*sinh(2*t)
```
(b)
```
>> syms s
>> f = 1/((s^2)*(s + 5));
>> ilaplace(f)
ans =
1/3*t-2/9*exp(-3/2*t)*sinh(3/2*t)
```
(c)
```
>>syms s
>> f=(3*s+1)/(s^2+2*s+9);
>> ilaplace(f)
ans =
3*exp(-t)*cos(2*2^(1/2)*t)-1/2*2^(1/2)*exp(-t)*sin(2*2^(1/2)*t)
```
(d)
```
>>syms s
>> f = (s -25)/(s*(s^2 + 3*s +25));
>> ilaplace(f)
ans =
5/4*exp(-3/2*t)*cos(1/2*71^(1/2)*t)+23/284*71^(1/2)*exp(-3/2*t)
*sin(1/2*71^(1/2)*t)-5/4
```

Example 2.36. *Find the inverse Laplace transform of the following function using* MATLAB.

$$G(s) = \frac{(s^2+9s+7)(s+7)}{(s+2)(s+3)(s^2+12s+150)}.$$

Solution:

% MATLAB Program

```
>> syms s    % tell MATLAB that "s" is a symbol.
>>G = (s^2 + 9*s +7)*(s + 7)/[(s + 2)*(s + 3)*(s^2 + 12*s + 150)]; % define
the function.
>>pretty(G) % the pretty function prints symbolic output
% in a format that resembles typeset mathematics.
          (s  + 9 s + 7) (s + 7)
      --------------------------------
          (s + 2) (s + 3) (s  + 12 s + 150)
>> g = ilaplace(G); % inverse Laplace transform
>>pretty(g)
                      44                 2915                        1/2
   - 7/26 exp(-2 t) + --- exp(-3 t) + ------ exp(-6 t) cos(114 t)
                      123                3198
         889                    1/2     1/2
    + ------- exp(-6 t) 114  sin(114  t)
        20254
```

Example 2.37. *Generate the transfer function using MATLAB.*

$$G(s) = \frac{3(s+9)\,(s+21)\,(s+57)}{s(s+30)\,(s^2+5s+35)\,(s^2+28s+42)}$$

using

(a) the ratio of factors

(b) the ratio of polynomials

Solution:

% MATLAB Program:

```
'a. The ratio of factors'
>>Gzpk = zpk([-9 -21 -57] , [0 -30 roots([1 5 35])'roots([1 28 42])'],3)
% zpk is used to create zero-pole-gain models or to convert TF or
% SS models to zero-pole-gain form.
'b. The ratio of polynomials'
>> Gp = tf(Gzpk)  % generate the transfer function
% Computer response:
ans =
```

(a) **The ratio of factors**

```
Zero/pole/gain:
          3 (s+9) (s+21) (s+57)
   -------------------------------------------
    s (s+30) (s+26.41) (s+1.59) (s^2 + 5s + 35)
ans =
```

(b) The ratio of polynomials

Transfer function:

```
        3 s^3 + 261 s^2 + 5697 s + 32319
--------------------------------------------------------
s^6 + 63 s^5 + 1207 s^4 + 7700 s^3 + 37170 s^2 + 44100 s
```

Example 2.38. *Generate the transfer function using MATLAB.*

$$G(s) = \frac{s^4 + 20s^3 + 27s^2 + 17s + 35}{s^5 + 8s^4 + 9s^3 + 20s^2 + 29s + 32}$$

using

(a) *the ratio of factors*

(b) *the ratio of polynomials*

Solution:

% MATLAB Program:

```
% a. the ratio of factors
>>Gtf = tf([1 20 27 17 35] , [1 8 9 20 29 32]) % generate the
% transfer function
% Computer response:
Transfer function:
    s^4 + 20 s^3 + 27 s^2 + 17 s + 35
    ---------------------------------
    s^4 + 8 s^3 + 9 s^2 + 20 s + 29
% b. the ratio of polynomials
>> Gzpk = zpk(Gtf)  % zpk is used to create zero-pole-gain models
% or to convert TF or SS models to zero-pole-gain form.
% Computer response:
Zero/pole/gain:
  (s+18.59) (s+1.623) (s^2 - 0.214s + 1.16)
  -------------------------------------------
  (s+7.042) (s+1.417) (s^2 - 0.4593s + 2.906)
```

2.23 SUMMARY

In this chapter the MATLAB environment which is an interactive environment for numeric computation, data analysis, and graphics was presented. Arithmetic operations, display formats, elementary built-in functions, arrays, scalars, vectors or matrices, operations with arrays including dot product, array multiplication, array division, inverse and transpose of a matrix, determinants, element by element operations, eigenvalues and eigenvectors, random number generating functions, polynomials, system of linear equation, script files, programming in MATLAB, the commands used for printing information and generating 2-D and 3-D plots, input/output in MATLAB was presented with illustrative examples. MATLAB's functions for symbolic mathematics were introduced. These functions are useful in performing symbolic operations and developing closed-form expressions for solutions to linear algebraic equations, ordinary differential equations and systems of equations. Symbolic mathematics for determining analytical expressions for the derivative and integral of an expression was also presented.

REFERENCES

Chapman, S.J., *MATLAB Programming for Engineers,* 2nd ed., Brooks/Cole, Thomson Learning, Pacific Grove, CA, 2002.

Etter, D.M., *Engineering Problem Solving with MATLAB,* Prentice-Hall, Englewood Cliffs, NJ, 1993.

Gilat, Amos., *MATLAB-An Introduction with Applications,* 2nd ed., Wiley, New York, 2005.

Hanselman, D., and Littlefield, B.R., *Mastering MATLAB 6,* Prentice Hall, Upper Saddle River, New Jersey, NJ, 2001.

Herniter, M.E., *Programming in MATLAB,* Brooks/Cole, Pacific Grove, CA, 2001.

Magrab, E.B., *An Engineers Guide to MATLAB,* Prentice Hall, Upper Saddle River, New Jersey, NJ, 2001.

Marchand, P., and Holland, O.T., *Graphics and GUIs with MATLAB,* 3rd ed, CRC Press, Boca Raton, FL, 2003.

Moler, C., *The Student Edition of MATLAB for MS-DOS Personal Computers with 3-1/ 2" Disks,* MATLAB Curriculum Series, The MathWorks, Inc., 2002.

Palm, W.J. III., *Introduction to MATLAB 7 for Engineers,* McGraw Hill, New York, NY, 2005.

Pratap, Rudra., *Getting Started with MATLAB— A Quick Introduction for Scientists and Engineers,* Oxford University Press, New York, NY, 2002.

Sigman, K., and Davis, T.A., *MATLAB Primer,* 6th ed, Chapman & Hall/CRCPress, Boca Raton, FL, 2002.

The MathWorks, Inc., *MATLAB: Application Program Interface Reference, Version 6,* The MathWorks, Inc., Natick, 2000.

The MathWorks, Inc., *MATLAB: Creating Graphical User Interfaces, Version 1,* The MathWorks, Inc., Natick, 2000.

The MathWorks, Inc., MATLAB: *Function Reference,* The MathWorks, Inc., Natick, 2000.

The MathWorks, Inc., *MATLAB: Release Notes for Release 12,* The MathWorks, Inc., Natick, 2000.

The MathWorks, Inc., *MATLAB: Symbolic Math Toolbox User's Guide, Version 2,* The MathWorks, Inc., Natick, 1993-1997.

The MathWorks, Inc., *MATLAB: Using MATLAB Graphics, Version 6,* The MathWorks, Inc., Natick, 2000.

PROBLEMS

P2.1 Compute the following quantity using MATLAB in the Command Window:

$$\frac{17\,[\sqrt{5}-1]}{[15^2-13^2]} + \frac{5^7\,\log_{10}(e^3)}{\pi\sqrt{121}} + \ln(e^4) + \sqrt{11}$$

P2.2 Compute the following quantity using MATLAB in the Command Window:

$$B = \frac{\tan x + \sin 2x}{\cos x} + \log |x^5 - x^2| + \cosh x - 2 \tanh x$$

for $x = \dfrac{5\pi}{6}$.

P2.3 Compute the following quantity using MATLAB in the Command Window:

$$x = a + \frac{ab}{c} \frac{(a+b)}{\sqrt{|ab|}} + c^a + \frac{\sqrt{14b}}{e^{3c}} + \ln(2) + \frac{\log_{10} c}{\log_{10}(a+b+c)} + 2 \sinh a - 3 \tanh b$$

for $a = 1$, $b = 2$ and $c = 1.8$.

P2.4 Use MATLAB to create

(a) a row and column vectors that has the elements: $11, -3, e^{7.8}, \ln(59), \tan(\pi/3), 5 \log_{10}(26)$.

(b) a row vector with 20 equally spaced elements in which the first element is 5.

(c) a column vector with 15 equally spaced elements in which the first element is -2.

P2.5 Enter the following matrix A in MATLAB and create:

$$A = \begin{bmatrix} 1 & 2 & 3 & 4 & 5 & 6 & 7 & 8 \\ 9 & 10 & 11 & 12 & 13 & 14 & 15 & 16 \\ 17 & 18 & 19 & 20 & 21 & 22 & 23 & 24 \\ 25 & 26 & 27 & 28 & 29 & 30 & 31 & 32 \\ 33 & 34 & 35 & 36 & 37 & 38 & 39 & 40 \end{bmatrix}$$

(a) a 4×5 matrix B from the 1st, 3rd, and the 5th rows, and the 1st, 2nd, 4th, and 8th columns of the matrix A.

(b) a 16 elements-row vector $\mathbf{C}$ from the elements of the 5th row, and the 4th and 6th columns of the matrix A.

P2.6 Given the function $y = \left(x^{\sqrt{2}+0.02} + e^x\right)^{1.8} \ln x$. Determine the value of y for the following values of x: 2, 3, 8, 10, -1, -3, -5, -6.2. Solve the problem using MATLAB by first creating a vector $\mathbf{x}$, and creating a vector $\mathbf{y}$, using element-by-element calculations.

P2.7 Define a and b as scalars, $a = 0.75$, and $b = 11.3$, and x, y and z as the vectors, $x = 2, 5, 1, 9$, $y = 0.2, 1.1, 1.8, 2$ and $z = -3, 2, 5, 4$. Use these variables to calculate A using element-by-element computations for the vectors with MATLAB.

$$A = \frac{x^{1.1}y^{-2}z^5}{(a+b)^{b/3}} + a \frac{\left(\dfrac{z}{x} + \dfrac{y}{2}\right)}{z^a}$$

P2.8 Enter the following three matrices in MATLAB and show that

$$A = \begin{bmatrix} 1 & 2 & 3 \\ -8 & 5 & 7 \\ -8 & 4 & 6 \end{bmatrix} B = \begin{bmatrix} 12 & -5 & 4 \\ 7 & 11 & 6 \\ 1 & 8 & 13 \end{bmatrix} C = \begin{bmatrix} 7 & 13 & 4 \\ -2 & 8 & -5 \\ 9 & -6 & 11 \end{bmatrix}$$

(a) $A + B = B + A$

(b) $A + (B + C) = (A + B)C$

(c) $7(A + C) = 7(A) + 7(C)$

(d) $A * (B + C) = A * B + A * C$

P2.9 Consider the function

$$H(s) = \frac{n(s)}{d(s)}$$

where $n(s) = s^4 + 6s^3 + 5s^2 + 4s + 3$

$d(s) = s^5 + 7s^4 + 6s^3 + 5s^2 + 4s + 7$

(a) Find $n(-10)$, $n(-5)$, $n(-3)$, and $n(-1)$

(b) Find $d(-10)$, $d(-5)$, $d(-3)$, and $d(-1)$

(c) Find $H(-10)$, $H(-5)$, $H(-3)$, and $H(-1)$

P2.10 Consider the polynomials

$$p_1(s) = s^3 + 5s^2 + 3s + 10$$
$$p_2(s) = s^4 + 7s^3 + 5s^2 + 8s + 15$$
$$p_3(s) = s^5 + 15s^4 + 10s^3 + 6s^2 + 3s + 9$$

Determine

(a) $p_1(2)$, $p_2(2)$, and $p_3(3)$

(b) $p_1(s)\, p_2(s)\, p_3(s)$

(c) $p_1(s)\, p_2(s)/p_3(s)$

P2.11 The following polynomials are given:

$$p_1(x) = x^5 + 2x^4 - 3x^3 + 7x^2 - 8x + 7$$
$$p_2(x) = x^4 + 3x^3 - 5x^2 + 9x + 11$$
$$p_3(x) = x^3 - 2x^2 - 3x + 9$$
$$p^4(x) = x^2 - 5x + 13$$
$$p^5(x) = x + 5$$

Use MATLAB functions with polynomial coefficient vectors to evaluate the expressions at $x = 2$.

P2.12 Determine the roots of the following polynomials:

(a) $p_1(x) = x^7 + 8x^6 + 5x^5 + 4x^4 + 3x^3 + 2x^2 + x + 1$

(b) $p_2(x) = x^6 - 7x^6 + 7x^5 + 15x^4 - 10x^3 - 8x^2 + 7x + 15$

(c) $p^3(x) = x^5 - 13x^4 + 10x^3 + 12x^2 + 8x - 15$

(d) $p_4(x) = x^4 + 7x^3 + 12x^2 - 25x + 8$

(e) $p_5(x) = x^3 + 15x^2 - 23x + 105$

(f) $p_6(x) = x^2 - 18x + 23$

(g) $p_7(x) = x + 7$

P2.13 Consider the two matrices

$$A = \begin{bmatrix} 1 & 0 & 2 \\ 2 & 5 & 4 \\ -1 & 8 & 7 \end{bmatrix} \text{ and } B = \begin{bmatrix} 7 & 8 & 2 \\ 3 & 5 & 9 \\ -1 & 3 & 1 \end{bmatrix}$$

Using MATLAB, determine the following:

(a) $A + B$

(b) AB

(c) $\mathbf{A}^2$

(d) $\mathbf{A}^{\mathrm{T}}$

(e) $\mathbf{B}^{-1}$

(f) $\mathbf{B}^{\mathrm{T}}\mathbf{A}^{\mathrm{T}}$

(g) $\mathbf{A}^2 + \mathbf{B}^2 - \mathbf{AB}$

(h) determinant of $\mathbf{A}$, determinant of $\mathbf{B}$ and determinant of $\mathbf{AB}$.

P2.14 Use MATLAB to define the following matrices:

$$\mathbf{A} = \begin{bmatrix} 2 & 1 \\ 0 & 5 \\ 7 & 4 \end{bmatrix} \qquad \mathbf{B} = \begin{bmatrix} 5 & 3 \\ -2 & -4 \end{bmatrix}$$

$$\mathbf{C} = \begin{bmatrix} 2 & 3 \\ -5 & -2 \\ 0 & 3 \end{bmatrix} \qquad \mathbf{D} = \begin{bmatrix} 1 & 2 \end{bmatrix}$$

Compute matrices and determinants if they exist.

(a) $(\mathbf{AC}^{\mathrm{T}})^{-1}$

(b) $|\mathbf{B}|$

(c) $|\mathbf{AC}^{\mathrm{T}}|$

(d) $(\mathbf{C}^{\mathrm{T}}\mathbf{A})^{-1}$

P2.15 Consider the two matrices

$$\mathbf{A} = \begin{bmatrix} 3 & 2\pi \\ 5j & 10 + \sqrt{2}j \end{bmatrix} \mathbf{B} = \begin{bmatrix} 7j & -15j \\ 2\pi & 18 \end{bmatrix}$$

Using MATLAB, determine the following:

(a) $\mathbf{A} + \mathbf{B}$

(b) $\mathbf{AB}$

(c) $\mathbf{A}^2$

(d) $\mathbf{A}^{\mathrm{T}}$

(e) $\mathbf{B}^{-1}$

(f) $\mathbf{B}^{\mathrm{T}}\mathbf{A}^{\mathrm{T}}$

(g) $\mathbf{A}^2 + \mathbf{B}^2 - \mathbf{AB}$

P2.16 Consider the two matrices

$$\mathbf{A} = \begin{bmatrix} 1 & 0 & 1 \\ 2 & 3 & 4 \\ -1 & 6 & 7 \end{bmatrix} \text{ and } \mathbf{B} = \begin{bmatrix} 7 & 4 & 2 \\ 3 & 5 & 6 \\ -1 & 2 & 1 \end{bmatrix}$$

Using MATLAB, determine the following:

(a) $\mathbf{A} + \mathbf{B}$

(b) $\mathbf{AB}$

(c) $\mathbf{A}^2$

(d) $\mathbf{A}^{\mathrm{T}}$

(e) $\mathbf{B}^{-1}$

(f) $\mathbf{B}^{\mathrm{T}}\mathbf{A}^{\mathrm{T}}$

(g) $\mathbf{A}^2 + \mathbf{B}^2 - \mathbf{AB}$

(h) det $\mathbf{A}$, det $\mathbf{B}$, and det of $\mathbf{AB}$.

P2.17 Find the inverse of the following Matrices:

(a) $\mathbf{A} = \begin{bmatrix} 3 & 2 & 1 \\ -1 & 5 & 4 \\ 5 & 7 & -9 \end{bmatrix}$ (b) $\mathbf{B} = \begin{bmatrix} 1 & 6 & 3 \\ -4 & -5 & 7 \\ 8 & 4 & 2 \end{bmatrix}$ (c) $\mathbf{C} = \begin{bmatrix} -1 & -2 & 5 \\ -4 & 7 & 2 \\ 7 & -8 & -1 \end{bmatrix}$

P2.18 Find the inverse of the following matrices using MATLAB.

(a) $\begin{bmatrix} 3 & 2 & 0 \\ 2 & -1 & 7 \\ 5 & 4 & 9 \end{bmatrix}$ (b) $\begin{bmatrix} -4 & 2 & 5 \\ 7 & -1 & 6 \\ 2 & 3 & 7 \end{bmatrix}$ (c) $\begin{bmatrix} -1 & 2 & -5 \\ 4 & 3 & 7 \\ 7 & -6 & 1 \end{bmatrix}$

(d) $\begin{bmatrix} 3 & 2 & 1 \\ -1 & 2 & 4 \\ 5 & 7 & -8 \end{bmatrix}$ (e) $\begin{bmatrix} 1 & 2 & 3 \\ -4 & -5 & 7 \\ 8 & 4 & 1 \end{bmatrix}$ (f) $\begin{bmatrix} -1 & -2 & 5 \\ -4 & 5 & 6 \\ 7 & 8 & -1 \end{bmatrix}$

P2.19 Determine the eigenvalues and eigenvectors of the following matrices using MATLAB.

$$A = \begin{bmatrix} 1 & -2 \\ 1 & 5 \end{bmatrix}, \quad B = \begin{bmatrix} 1 & 5 \\ -2 & 7 \end{bmatrix}$$

P2.20 If $A = \begin{bmatrix} 4 & 6 & 2 \\ 5 & 6 & 7 \\ 10 & 5 & 8 \end{bmatrix}$

Use MATLAB to determine the following:

(a) the three eigenvalues of $\mathbf{A}$

(b) the eigenvectors of $\mathbf{A}$

(c) Show that $\mathbf{AQ} = \mathbf{Qd}$ where Q is the matrix containing the eigenvectors as columns and $\mathbf{d}$ is the matrix containing the corresponding eigenvalues on the main diagonal and zeros elsewhere.

P2.21 Determine eigenvalues and eigenvector of $\mathbf{A}$ using MATLAB.

(a) $\mathbf{A} = \begin{bmatrix} 0.5 & -0.8 \\ 0.75 & 1.0 \end{bmatrix}$ (b) $\mathbf{A} = \begin{bmatrix} 8 & 3 \\ -3 & 4 \end{bmatrix}$

P2.22 Determine the eigenvalues and eigenvectors of the following matrices using MATLAB.

(a) $\mathbf{A} = \begin{bmatrix} 1 & -2 \\ 1 & 3 \end{bmatrix}$ (b) $\mathbf{A} = \begin{bmatrix} 1 & 5 \\ -2 & 4 \end{bmatrix}$

(c) $\mathbf{A} = \begin{bmatrix} 4 & -1 & 5 \\ 2 & 1 & 3 \\ 6 & -7 & 9 \end{bmatrix}$ (d) $\mathbf{A} = \begin{bmatrix} 3 & 5 & 7 \\ 2 & 4 & 8 \\ 5 & 6 & 10 \end{bmatrix}$

(e) $\mathbf{A} = \begin{bmatrix} 3 & 0 & 2 & 1 \\ 1 & 2 & 5 & 4 \\ 7 & -1 & 2 & 6 \\ 1 & -2 & 3 & 4 \end{bmatrix}$ (f) $\mathbf{A} = \begin{bmatrix} 1 & 3 & 5 & 7 \\ 2 & -1 & -2 & 4 \\ 3 & 2 & 1 & 1 \\ 4 & 1 & 0 & 6 \end{bmatrix}$

P2.23 Determine the eigenvalues and eigenvectors of $A * B$ using MATLAB.

$$A = \begin{bmatrix} 3 & -1 & 2 & 1 \\ 1 & 2 & 7 & 4 \\ 7 & -1 & 8 & 6 \\ 1 & -2 & 3 & 4 \end{bmatrix} \qquad B = \begin{bmatrix} 1 & 2 & 5 & 7 \\ 2 & -1 & -2 & 4 \\ 3 & 2 & 5 & 1 \\ 4 & 1 & -3 & 6 \end{bmatrix}$$

P2.24 Determine the eigenvalues and eigenvectors of the following matrices using MATLAB.

(a) $\mathbf{A} = \begin{bmatrix} 1 & -2 \\ 1 & 3 \end{bmatrix}$
(b) $\mathbf{A} = \begin{bmatrix} 1 & 5 \\ -2 & 4 \end{bmatrix}$

(c) $\mathbf{A} = \begin{bmatrix} 4 & -1 & 5 \\ 2 & 1 & 3 \\ 6 & -7 & 9 \end{bmatrix}$
(d) $\mathbf{A} = \begin{bmatrix} 3 & 5 & 7 \\ 2 & 4 & 8 \\ 5 & 6 & 10 \end{bmatrix}$

(e) $\mathbf{A} = \begin{bmatrix} 3 & 0 & 2 & 1 \\ 1 & 2 & 5 & 4 \\ 7 & -1 & 2 & 6 \\ 1 & -2 & 3 & 4 \end{bmatrix}$
(f) $\mathbf{A} = \begin{bmatrix} 1 & 3 & 5 & 7 \\ 2 & -1 & -2 & 4 \\ 3 & 2 & 1 & 1 \\ 4 & 1 & 0 & 6 \end{bmatrix}$

P2.25 Determine the eigenvalues and eigenvectors of **A** and **B** using MATLAB

(a) $\mathbf{A} = \begin{bmatrix} 4 & 5 & -3 \\ -1 & 2 & 3 \\ 2 & 5 & 7 \end{bmatrix} \qquad \mathbf{B} = \begin{bmatrix} 1 & 2 & 3 \\ 8 & 9 & 6 \\ 5 & 3 & -1 \end{bmatrix}$

P2.26 Determine the eigenvalues and eigenvectors of $A = a*b$ using MATLAB.

$$\mathbf{a} = \begin{bmatrix} 6 & -3 & 4 & 1 \\ 0 & 4 & 2 & 6 \\ 1 & 3 & 8 & 5 \\ 2 & 2 & 1 & 4 \end{bmatrix} \qquad \mathbf{b} = \begin{bmatrix} 0 & 1 & 2 & 3 \\ 4 & 5 & 6 & -1 \\ 1 & 5 & 4 & 2 \\ 2 & -3 & 6 & 7 \end{bmatrix}$$

P2.27 Determine the values of x, y, and z for the following set of linear algebraic equations:

$$x_2 - 3x_3 = -7$$
$$2x_1 + 3x_2 - x_3 = 9$$
$$4x_1 + 5x_2 - 2x_3 = 15$$

P2.28 Determine the values of x, y, and z for the following set of linear algebraic equations:

(a) $2x + y - 3z = 11$
$4x - 2y + 3z = 8$
$-2x + 2y - z = -6$

(b) $2x - y = 10$
$-x + 2y - z = 0$
$-y + z = -50$

P2.29 Solve the following set of equations using MATLAB.

(a) $2x_1 + x_2 + x_3 - x_4 = 12$
$x_1 + 5x_2 - 5x_3 + 6x_4 = 35$
$-7x_1 + 3x_2 - 7x_3 - 5x_4 = 7$
$x_1 - 5x_2 + 2x_3 + 7x_4 = 21$

(b) $x_1 - x_2 + 3x_3 + 5x_4 = 7$

$\quad\quad 2x_1 + x_2 - x_3 + x_4 = 6$

$\quad\quad -x_1 - x_2 - 2x_3 + 2x_4 = 5$

$\quad\quad x_1 + x_2 - x_3 + 5x_4 = 4$

P2.30 Solve the following set of equations using MATLAB.

(a) $\quad 2x_1 + x_2 + x_3 - x_4 = 10$

$\quad\quad x_1 + 5x_2 - 5x_3 + 6x_4 = 25$

$\quad\quad -7x_1 + 3x_2 - 7x_3 - 5x_4 = 5$

$\quad\quad x_1 - 5x_2 + 2x_3 + 7x_4 = 11$

(b) $x_1 - x_2 + 3x_3 + 5x_4 = 5$

$\quad\quad 2x_1 + x_2 - x_3 + x_4 = 4$

$\quad\quad -x_1 - x_2 + 2x_3 + 2x_4 = 3$

$\quad\quad x_1 + x_2 - x_3 + 5x_4 = 1$

P2.31 Solve the following set of equations using MATLAB.

(a) $\quad x_1 + 2x_2 + 3x_3 + 5x_4 = 21$

$\quad\quad -2x_1 + 5x_2 + 7x_3 - 9x_4 = 17$

$\quad\quad 5x_1 + 7x_2 + 2x_3 - 5x_4 = 23$

$\quad\quad -x_1 - 3x_2 - 7x_3 + 7x_4 = 26$

(b) $\quad x_1 + 2x_2 + 3x_3 + 4x_4 = 9$

$\quad\quad 2x_1 - 2x_2 - x_3 + x_4 = -5$

$\quad\quad x_1 - 3x_2 + 4x_3 - 4x_4 = 7$

$\quad\quad 2x_1 + 2x_2 - 3x_3 + 4x_4 = -6$

P2.32 Generate a plot of

$$y(x) = e^{-0.7x} \sin \omega x$$

where $\omega = 15$ rad/s, and $0 \le x \le 15$. Use the colon notation to generate the x vector in increments of 0.1.

P2.33 Plot the following functions using MATLAB.

(a) $r^2 = 5 \cos 3t$ $\quad\quad\quad\quad\quad 0 \le t \le 2\pi$

(b) $r^2 = 5 \cos 3t$ $\quad\quad\quad\quad\quad 0 \le t \le 2\pi$

$\quad\quad x = r \cos t, \; y = r \sin t$

(c) $y_1 = e^{-2x} \cos x$ $\quad\quad\quad\quad 0 \le x \le 20$

$\quad\quad y_2 = e^{2x}$

(d) $y = \cos (x)/x$ $\quad\quad\quad\quad\quad -5\pi \le x \le 5\pi$

(e) $f = e^{-3t/5} \cos t$ $\quad\quad\quad\quad 0 \le t \le 2\pi$

(f) $z = -(1/3) x^2 + 2xy + y^2$ $\quad |x| \le 7, \; |y| \le 7$

P2.34 Use MATLAB for plotting 3-D data for the following functions:

(a) $z = \cos x \cos y \, e^{-\sqrt{\frac{x^2 + y^2}{5}}}$ $\quad |x| \le 7, \, |y| \le 7$

(b) Discrete data plots with stems

$\quad\quad x = t, \; y = t \cos (t)$

$\quad\quad z = et/5 - 2$ $\quad\quad\quad 0 \le x \le 5\pi$

(c) An ellipsoid of radii $rx = 1$, $ry = 2.5$ and $rz = 0.7$ centered at the origin

(d) A cylinder generated by

$$r = \sin(5\pi z) + 3 \quad 0 \le z \le 1$$
$$0 \le \theta \le 2\pi$$

P2.35 Obtain the plot of the points for $0 \le t \le 6\pi$ when the coordinates x, y, and z are given as a function of the parameter t as follows:

$$x = \sqrt{t} \sin(3t)$$

$$y = \sqrt{t} \cos(3t)$$

$$z = 0.8t$$

P2.36 Obtain the mesh and surface plots for the function $z = \dfrac{2xy^2}{x^2 + y^2}$ over the domain $-2 \le x \le 6$ and $2 \le y \le 8$.

P2.37 Plot the function $z = 2^{-1.5\sqrt{x^2 + y^2}} \sin(x) \cos(0.5y)$ over the domain $-4 \le x \le 4$ and $-4 \le y \le 4$.

(a) Mesh plot

(b) Surface plot

(c) Mesh curtain plot

(d) Mesh and contour plot

(e) Surface and contour plot

(f) Surface plot with lighting

(g) Waterfall plot

(h) 3-D contour plot

(i) 2-D contour plot

P2.38 Plot the function $y = |x| \cos(x)$ for $-200 \le x \le 200$.

P2.39 Plot the following functions on the same plot for $0 \le x \le 2\pi$ using the plot function:

(a) $\sin^2(x)$

(b) $\cos^2 x$

(c) $\cos(x)$

P2.40 (a) Generate an overlay plot for plotting three lines

$$y_1 = \sin t$$
$$y_2 = t$$

$$y_3 = t - \frac{t^3}{3!} + \frac{t^5}{5!} + \frac{t^7}{7!} \quad 0 \le t \le 2\pi$$

Use   (i) the plot command

(ii) the hold command

(iii) the line command

(b) Use the functions for plotting x-y data given in Table 6.5(b) for plotting the following functions.

(i) $f(t) = t \cos t$ $\qquad\qquad\qquad\qquad\qquad$ $0 \le t \le 10\pi$

(ii) $x = e^t$

$\qquad$ $y = 100 + e^{3t}$ $\qquad\qquad\qquad\qquad$ $0 \le t \le 2\pi$

P2.41 (a) Plot the parametric space curve of

$$x(t) = t$$
$$y(t) = t^2$$
$$z(t) = t^3 \qquad 0 \le t \le 3.0$$

(b) $z = \dfrac{-7}{1 + x^2 + y^2}$ $\qquad$ $|x| \le 10,\ |y| \le 10$

P2.42 Perform the following symbolic operations using MATLAB. Consider the given symbolic expressions have been defined.

$$S_1 = \text{'2/(x − 5)'};$$
$$S_2 = \text{'x ∧ 5 + 9* x − 15'};$$
$$S_3 = \text{'(x ∧ 3 + 2* x + 9)* (x* x − 5)'};$$

(a) $S_1 S_2/S_3$ (b) $S_1/S_2 S_3$ (c) $S_1/(S_2)^2$ (d) $S_1 S_3/S_2$ (e) $(S_2)^2/(S_1 S_3)$

P2.43 Solve the following equations using symbolic mathematics.

(a) $x^2 + 9 = 0$

(b) $x^2 + 5x - 8 = 0$

(c) $x^3 + 11x^2 - 7x + 8 = 0$

(d) $x^4 + 11x^3 + 7x^2 - 19x + 28 = 0$

(e) $x^7 - 8x^5 + 7x^4 + 5x^3 - 8x + 9 = 0$

P2.44 Determine the values of x, y, and z for the following set of linear algebraic equations:

$$2x + y - 3z = 11$$
$$4x - 2y + 3z = 8$$
$$-2x + 2y - z = -6$$

P2.45 Determine the values of x, y, and z for the following set of linear algebraic equations:

$$2x - y = 10$$
$$-x + 2y - z = 0$$
$$-y + z = -50$$

P2.46 Determine the solutions of the following first-order ordinary differential equations using MATLAB's symbolic mathematics.

(a) $y' = 8x^2 + 5$ with initial condition $y(2) = 0.5$.

(b) $y' = 5x \sin^2(y)$ with initial condition $y(0) = \pi/5$.

(c) $y' = 7x \cos^2(y)$ with initial condition $y(0) = 2$.

(d) $y' = -5x + y$ with initial condition $y(0) = 3$.

(e) $y' = 3y + e^{-5x}$ with initial condition $y(0) = 2$.

P2.47 (a) Given the differential equation

$$\frac{dx^2}{dt^2} + 7\frac{dx}{dt} + 5x = 8u(t) \quad t \geq 0$$

Using MATLAB program, find

    (i) $x(t)$ when all the initial conditions are zero

    (ii) $x(t)$ when $x(0) = 1$ and $\dot{x}(0) = 3$.

(b) Given the differential equation

$$\frac{d^2x}{dt^2} + 12\frac{dx}{dt} + 15x = 35 \qquad t \geq 0$$

Using MATLAB program, find

    (i) $x(t)$ when all the initial conditions are zero

    (ii) $x(t)$ when $x(0) = 0$ and $\dot{x}(0) = 1$.

    (iii) For the following differential equation, use MATLAB to find $x(t)$ when $x(0) = -1$ and $\dot{x}(0) = 1$

$$\frac{d^2x}{dt^2} + 8\frac{dx}{dt} - 4x = 18\ u(t)$$

(c) For the following differential equation, use MATLAB to find $x(t)$ when $x(0) = -1$ and $\dot{x}(0) = 1$

$$\frac{d^2x}{dt^2} + 15\frac{dx}{dt} + 8x = -9\ u(t)$$

(d) For the following differential equation, use MATLAB to find $x(t)$ when $x(0) = -1$ and $\dot{x}(0) = 1$

$$\frac{d^2x}{dt^2} - 19\frac{dx}{dt} + 9x = -3\ u(t)$$

P2.48 For the following differential equations, use MATLAB to find $x(t)$ when (a) all the initial conditions are zero, (b) $x(t)$ when $x(0) = 1$ and $\dot{x}(0) = -1$.

(a) $\dfrac{d^2x}{dt^2} + 10\dfrac{dx}{dt} + 5x = 11$

(b) $\dfrac{d^2x}{dt^2} - 7\dfrac{dx}{dt} - 3x = 5$

(c) $\dfrac{d^2x}{dt^2} + 3\dfrac{dx}{dt} + 7x = -15$

(d) $\dfrac{d^2x}{dt^2} + \dfrac{dx}{dt} + 7x = 26$

P2.49 Obtain the first and second derivatives of the following functions using MATLAB's symbolic mathematics.

(a) $F(x) = x^5 - 8x^4 + 5x^3 - 7x^2 + 11x - 9$

(b) $F(x) = (x^3 + 3x - 8)(x^2 + 21)$

(c) $F(x) = (3x^3 - 8x^2 + 5x + 9)/(x + 2)$

(d) $F(x) = (x^5 - 3x^4 + 5x^3 + 8x^2 - 13)^2$

(e) $F(x) = (x^2 + 8x - 11)/(x^7 - 7x^6 + 5x^3 + 9x - 17)$

P2.50 Determine the values of the following integrals using MATLAB's symbolic functions.

(a) $\int 5x^7 - lx^5 + 3x^3 - 8x^2 + 7$

(b) $\int \sqrt{x} \cos x$

(c) $\int x^{2/3} \sin^2 2x$

(d) $\int_{0.2}^{1.8} x^2 \sin x \, dx$

(e) $\int_{-1}^{0.2} |x| \, dx$

P2.51 Given the differential equation

$$\frac{d^2x}{dt^2} + 3\frac{dx}{dt} + x = 98 \qquad\qquad t \geq 0$$

Using MATLAB program, find

(a) $x(t)$ when all the initial conditions are zero

(b) $x(t)$ when $x(0) = 0$ and $\dot{x}(0) = 2$

P2.52 Determine the inverse of the following matrix using MATLAB.

$$A = \begin{bmatrix} 3s & 2 & 0 \\ 7s & -s & -5 \\ 3 & 0 & -3s \end{bmatrix}$$

P2.53 Expand the following function $F(s)$ into partial fractions with MATLAB:

$$F(s) = \frac{5s^3 + 7s^2 + 8s + 30}{s^4 + 15s^3 + 62s^2 + 85s + 25}$$

P2.54 Determine the Laplace transform of the following time functions using MATLAB.

(a) $f(t) = u(t + 9)$

(b) $f(t) = e^{5t}$

(c) $f(t) = (5t + 7)$

(d) $f(t) = 5u(t) + 8e^{7t} - 12e^{-8t}$

(e) $f(t) = e^{-t} + 9t^3 - 7t^{-2} + 8$

(f) $f(t) = 7t^4 + 5t^2 - e^{-7t}$

(g) $f(t) = 9ut + 5e^{-3t}$

P2.55 Determine the inverse Laplace transform of the following rotational function using MATLAB.

$$F(s) = \frac{7}{s^2 + 5s + 6} = \frac{7}{(s + 2)(s + 3)}$$

P2.56 Determine the inverse transform of the following function having complex poles.

$$F(s) = \frac{15}{(s^3 + 5s^2 + 11s + 10)}$$

P2.57 Determine the inverse Laplace transform of the following functions using MATLAB:

$(a)\ F(s) = \dfrac{s}{s(s+2)\,(s+3)\,(s+5)}$

$(b)\ F(s) = \dfrac{1}{s^2(s+7)}$

$(c)\ F(s) = \dfrac{5s+9}{(s^3+8s+5)}$

$(d)\ F(s) = \dfrac{s-28}{s(s^2+9s+33)}$

CHAPTER 3

MATLAB Tutorial

3.1 INTRODUCTION

MATLAB has an excellent collection of commands and functions that are useful for solving vibration analysis problems. The problems presented in this chapter are basic linear vibrating systems and are normally presented in introductory mechanical vibrations courses. The application of MATLAB to the analysis vibrating systems is presented in this chapter with a number of illustrative examples. The MATLAB computational approach to the transient response analysis to the simple inputs is presented.

3.2 EXAMPLE PROBLEMS AND SOLUTIONS

Example 3.1. *Write a MATLAB script for plotting*

(a) the non-dimensional response magnitude for a system with harmonically moving base shown in Fig. E3.1.

(b) the response phase angle for system with harmonically moving base.

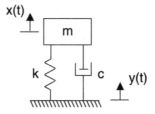

Fig. E3.1 Single degree of freedom system with moving base.

Solution:

The magnitude of the frequency response is given as

$$|G(i\omega)| = \frac{1}{\left[\left[\left(1 - \frac{\omega}{\omega_n}\right)^2\right]^2 + \left(2\zeta\frac{\omega}{\omega_n}\right)^2\right]^{1/2}}$$

150

The magnitude of $X(i\omega)$ is given as

$$|X(i\omega)| = \left[1+\left(\frac{2\zeta\omega}{\omega_n}\right)^2\right]^{1/2} |G(i\omega)| A$$

where $y(t) = \text{Re } A^{i\omega t}$

$x(t) = X(i\omega) e^{i\omega t}$

The phase angle ϕ is given as

$$\phi(\omega) = \tan^{-1}\left[\frac{2\zeta\left(\dfrac{\omega}{\omega_n}\right)^3}{1-\left(\dfrac{\omega}{\omega_n}\right)^2+\left(\dfrac{2\zeta\omega}{\omega_n}\right)^2}\right]$$

The frequency ratio

$$r = \frac{\omega}{\omega_n}$$

The non-dimensional response magnitude is given as the transmissibility

$$\frac{|X(i\omega)|}{A} = \left[\frac{1+\left(\dfrac{2\zeta\omega}{\omega_n}\right)^2}{1-\left(\dfrac{\omega}{\omega_n}\right)^2+\left(\dfrac{2\zeta\omega}{\omega_n}\right)^2}\right]$$

Based on these equations MATLAB script is written as follows:

```
zeta= [0.05; 0.1; 0.15; 0.25; 0.5; 1.25; 1.5]; % damping factors
r= [0:0.01:3]; %frequency ratio
for k=1: length (zeta)
    G(k,:)=sqrt((1+(2*zeta(k)*r).^2)./((1-r.^2).^2+(2*zeta(k)*r).^2));
    phi(k,:)=atan2(2*zeta(k)*r.^3,1-r.^2+(2*zeta(k)*r).^2);
end
figure (1)
plot(r, G)
xlabel ('\omega/\omega_n')
ylabel ('|x (i\omega)|/A')
grid
legend ('\zeta_1=0.05','\zeta_2=0.1','\zeta_3=0.15','\zeta_4
    =0.25','\zeta_5=0.5','\zeta_6=1.25','\zeta_7=1.5')
figure (2)
plot(r, phi)
xlabel ('\omega/\omega_n')
ylabel ('\phi (\omega)')
grid
ha=gca;
set (ha,'ytick',[0:pi/2:pi])
```

```
set(ha,'yticklabel',{[];'pi/2';'p'})
legend('\zeta_1=0.05','\zeta_2=0.1','\zeta_3=0.15','\zeta_
4=0.25','\zeta_5=0.5','\zeta_6=1.25','\zeta_7=1.5')
```

The output of this program is shown in Fig. E3.1(a) and (b).

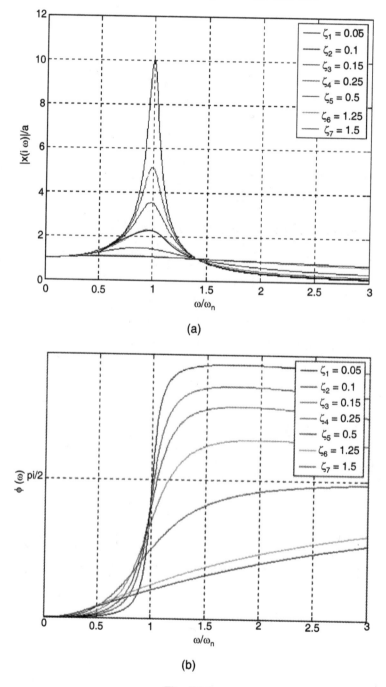

(a)

(b)

Fig. E3.1

Example 3.2. *An analytical expression for the response of an damped single degree of freedom system (Fig. E3.2) to given initial displacement and velocity is given by*

$$x(t) = C\, e^{-\zeta \omega_n t}\, \cos\,(\omega_d t - \phi)$$

where C and ϕ represent the amplitude and phase angle of the response, respectively having the values

$$C = \sqrt{x_0^2 + \left(\frac{\zeta \omega_n x_0 + v_0}{\omega_d}\right)^2}\ ,\ \phi = tan^{-1}\left(\frac{\zeta \omega_n x_0 + v_0}{\omega_d x_0}\right)\ and\ \ \omega_d = \sqrt{1 - \zeta^2}\ \omega_n$$

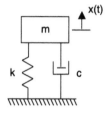

Fig. E3.2

Plot the response of the system using MATLAB for ω_n = 5rad/s, ζ = 0.05, 0.1, 0.2 *subjected to the initial conditions* x(0) = 0, $\dot{x}(0) = v_0$ = 60 cm/s.

Solution:

```
clear
clf
wn=5; % Natural frequency
zeta=[0.05;0.1;0.2]; % Damping ratio
x0=0; % Initial displacement
v0=60; % Initial velocity
t0=0; % Initial time
deltat=0.01; % Time step
tf=6; % Final time
t=[t0:deltat:tf];
for i=1:length(zeta),
    wd=sqrt(1-zeta(i)^2)*wn; % Damped frequency
    x=exp(-zeta(i)*wn*t).*(((zeta(i)*wn*x0+v0)/wd)*sin(wd*t)
    + x0*cos(wd*t));
    plot(t,x)
        hold on
end
title('Response to initial excitations')
xlabel('t[s]')
ylabel('x(t)')
grid
```

The output of this program is as follows:

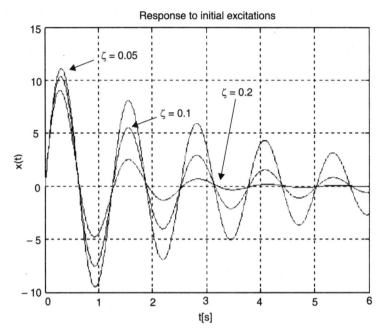

Fig. E3.2(a)

Example 3.3. *Plot the response of the system in Problem E3.2 using MATLAB for* $\omega_n = 5$ *rad/sec,* $\zeta = 1.3, 1.5, 2.0$ *subjected to the initial conditions* $x(0) = 0,\ \dot{x}(0) = v_0 = 60$ *cm/s.*

Solution:

Changing the program slightly, with zeta = [1.3, 1.5, 2.0] in E3.2, we obtain Fig. E3.3.

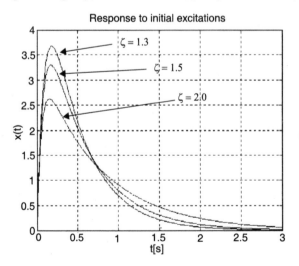

Fig. E3.3

Example 3.4. *Plot the response of the system in Problem E3.2 using MATLAB for $\omega_n = 5$ rad/sec and $\zeta = 1.0$ subjected to the initial conditions $x(0) = 0$, $\dot{x}(0) = v_0 = 60$ cm/s.*

Solution:

The solution obtained is shown in Fig. E3.4.

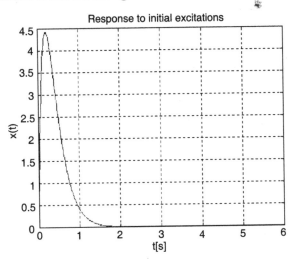

Fig. E3.4

Example 3.5. *Write MATLAB script for plotting the magnitude of the frequency response of a system with rotating unbalanced masses as shown in Fig. E3.5.*

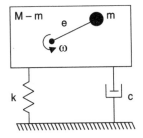

Fig. E3.5 Single degree of freedom system with rotating eccentric mass.

Hint: The magnitude of the frequency response is given as

$$|G(i\omega)| = \frac{1}{\left[\left(1-\left(\frac{\omega}{\omega_n}\right)^2\right)^2 + \left(2\zeta\frac{\omega}{\omega_n}\right)^2\right]^{1/2}}$$

Solution:

The magnitude of the frequency response is given as

$$|G(i\omega)| = \frac{1}{\left[\left(1-\left(\frac{\omega}{\omega_n}\right)^2\right)^2 + \left(2\zeta\frac{\omega}{\omega_n}\right)^2\right]^{1/2}}$$

for ζ = 0.05, 0.01, 0.15, 0.20, 0.20, 0.25, 0.5, 0.75, 1.0, 1.25, 1.5.

$r = \omega/\omega_n$ = 0 to 3 in steps of 0.01.

Following MATLAB program is developed:

```
zeta=[0.05;0.1;0.15;0.25;0.5;1;1.25;1.5]; % Damping factors
r=[0:0.01:3]; % Frequency ratios
for k=1:length(zeta),
    G=(r.^2)./sqrt((1-r.^2).^2+(2*zeta(k)*r).^2);
    plot(r,G)
    hold on
end
xlabel('\omega/\omega_n')
ylabel('({\omega/\omega_n})^2|G(I\omega)|')
grid
```

Fig. E3.5 (a) shows the output of the program

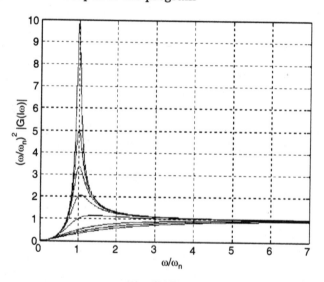

Fig. E3.5(a)

For showing legends on the curves, gtext command can be employed.

Example 3.6. *A single degree of freedom spring-mass system subjected to coulomb damping is shown in Fig. E3.6.*

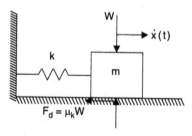

Fig. E3.6

The parameters of the system have the values $m = 600$ kg, $k = 20 \times 10^4$ N/m, $\mu_s = 0.15$ and $\mu_k = 0.10$. The initial conditions are $x(0) = x_0 = 1.5$ cm, $\dot{x}(0) = 0$. Plot the response $x(t)$ versus t using MATLAB.

The magnitude of the average response value f_d is given as

$$f_d = \frac{F_d}{k} = \frac{\mu_k \, mg}{k}$$

If n denotes the half-cycle just prior to the cessation of motion, then n is the smallest integer satisfying the inequality

$$x_0 - (2n - 1)f_d < \left(1 + \frac{\mu_s}{\mu_f}\right) f_d$$

where μ_s = static coefficient of friction

μ_k = kinetic coefficient of friction

Solution:

The following MATLAB program can be developed:

```
m=600; % Mass
k=200000; % Stiffness
mus=0.15; % Static friction coefficient
muk=0.10; % Kinetic friction coefficient
x0=1.5; % Initial displacement
t0=0;
deltat=0.005; % Time increment
wn=sqrt(k/m); % Natural frequency
fd=100*muk*m*9.81/k;
N=ceil(0.5*((x0-(1+mus/muk)*fd)/fd+1)); % Half cycles
t=[];
x=[];
if N>0
      for n=1:N,
          t1=[t0:deltat:t0+pi/wn];
          x1=(x0-(2*n-1)*fd)*cos(wn*t1)+fd*(-1)^(n+1);
          t=[t t1];
          x=[x x1];
          t0=t0+pi/wn;
      end
end
plot(t,x,t,fd*ones(length(t)),'--',t,-fd*ones(length(t)),'--')
title('Response to initial excitations')
xlabel('t[s]')
ylabel('x(t)[cm]')
grid
```

The output is shown in Fig. E3.6(a).

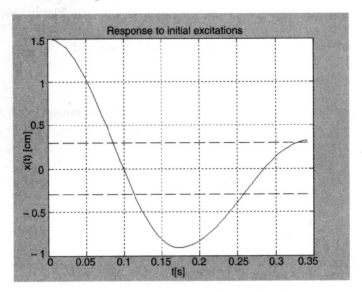

Fig. E3.6(a)

Example 3.7. *Write a MATLAB script for obtaining the response of a viscosity damped single degree of freedom system to the force $F(t) = F_0 e^{-\alpha t} u(t)$ by means of the convolution integral. The pulse is rectangular as shown in Fig. E3.7 with $T = 0.1$ seconds.*

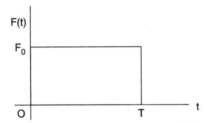

Fig. E3.7 Rectangular pulse.

Use the sampling period of $T = 0.001$ s and the number of sampling times $n = 300$. The parameters of the system are given as $m = 25$ kg, $c = 30$ Ns/m, $k = 6000$ N/m, $F_0 = 300$ N, and $\alpha = 1$. The impulse response of a mass-damper spring system is given by

$$g(t) = \frac{1}{m\omega_d} e^{-\zeta\omega_n t} \sin \omega_d t \, u(t)$$

Solution:

```
m=25; % mass
c=30;% damping
k=6000; % stiffness
F0=300; % Force amplitude
T=0.1;
wn=sqrt(k/m);% Natural frequency
```

```
zeta=c/(2*sqrt(m*k));%damping factor
Ts=0.001;% sampling period
N=301;% sampling times
wd=wn*sqrt(1-zeta^2);% damped frequency
for n=1:N,
   if n<=T/Ts+1; F(n)=F0; else F(n)=0; end      %force
end
n=[1:N];
g=Ts*exp(-(n-1)*zeta*wn*Ts).*sin((n-1)*wd*Ts)/(m*wd);
% discrete-time impulse response
c0=conv(F,g);%convolution sum
c=c0(1:N); % plot to N samples
n=[0:N-1];
axes('position',[0.1 0.2 0.8 0.7])
plot(n,c,'.')
title('Response to Rectangular pulse')
xlabel('n')
ylabel('x(n) m');
grid
```

The output is shown in Fig. E3.7 (*a*).

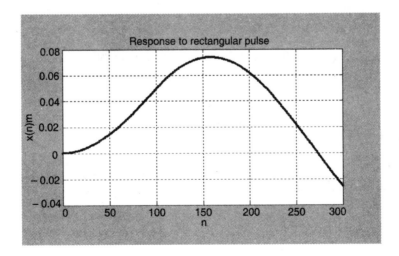

Fig. E3.7(a)

Example 3.8. *A simplified single degree of freedom model of an automobile suspension system is shown in Fig. E3.8. The automobile is traveling over a rough road at a constant horizontal speed when it encounters a bump in the road of the shape shown in Fig. E3.8(a), (b). The velocity of the automobile is 20 m/s, m = 1500 kg, k = 150,000 N/m, and ζ = 0.10. Determine the response of the automobile.*

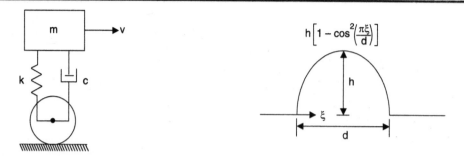

Fig. E3.8 Simplified single degree of pulse model for bump.

Fig. E3.8(a) Versed sine Freedom automobile model.

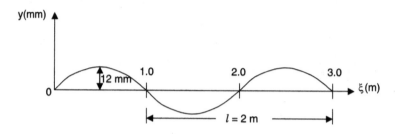

Fig. E3.8(b) Road contour.

$$y(\xi) = h \left[1 - \cos^2 \left(\frac{\pi \xi}{d} \right) \right] [1 - u(\xi - d)]$$

Here $h = 0.012$ m, $d = 1.0$ m and for constants automobile speed, $\xi = vt$. The vertical displacement of the automobile wheels is given by

$$y(t) = h \left[1 - \cos^2 \left(\frac{\pi v}{d} t \right) \right] \left[1 - u \left(t - \frac{d}{v} \right) \right]$$

The system response as per convolution integral is

$$x(t) = - m_{eq} \int_0^t [2\zeta \omega_n \dot{y}(\tau) + \omega_n^2 y(\tau)] h(t - \tau) \, d\tau$$

The wheel velocity becomes

$$\dot{y}(t) = 2 \left(\frac{\pi v}{d} \right) \sin \left(\frac{2\pi v}{d} t \right) \left[1 - u \left(t - \frac{d}{v} \right) \right]$$

Solution:

MATLAB program for this is given below:

```
% Simplified one-degree-of-freedom model of vehicle suspension system
% Vehicle encounters bump in road modelled as a versed sinusoidal pulse
% y(t)=h(1-(cos(pi*v*t/ t0))^2)*(u(t)-u(t-d/v))
%convolution integral is used to evaluate system response
syms t tau
% input parameters
digits(10)
```

```
format short e
m=1500;
k=150000;
zeta=0.10;
hb=0.012;
d=1.0;
v=20;
% system parameters and constants
omega_n=sqrt(k/m); % Natural frequency
omega_d=omega_n*sqrt(1-zeta^2); % damped natural frequency
c1=pi/d;
% wheel displacement and velocity
% MATLAB  'Heaviside' for the unit step function
y=hb*(1-cos(c1*v*t)^2)*(1-sym('Heaviside(t-0.04)'));
ydot=hb*c1*sin(2*c1*v*tau)*(1-sym('Heaviside(tau-0.04)'))
%convolution integral evaluation
h=exp(-zeta*omega_n*(t-tau)).*sin(omega_d*(t-tau))/(m*omega_d);
g1=-2*zeta*m*omega_n*ydot*h;
g2=-omega_n^2*m*y*h;
g1a=vpa(g1,5);
g2a=vpa(g2,5);
I1=int(g1a,tau,0,t);
I1a=vpa(I1,5);
I2=int(g2a,tau,0,t);
I2a=vpa(I2,5);
x1=I1a+I2a;
x=vpa(x1,5);
vel=diff(x);
acc=diff(vel);
time=linspace(0,0.3,50);
for i=1:50
   x1=subs(x,t,time(i));
   xa(i)=vpa(x1);
end
xp=double(xa);
plot(time,xp,'-');
grid;
xlabel('time(sec)')
ylabel('x(t)  [m]')
```

The output of this MATLAB program is given in Fig. E3.8(*c*)

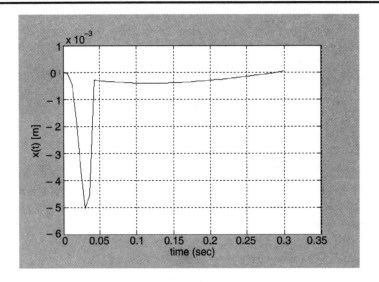

Fig. E3.8(c)

Example 3.9. *Fig. E3.9 shows two disks of mass polar moments of inertia I_1 and I_2 mounted on a circular shaft with torsional stiffnesses G_{J1} and G_{J2}. Neglect the mass of the shaft.*

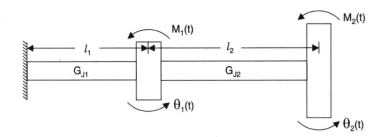

Fig. E3.9

(a) *Obtain the differential equations of motion for the angular displacements of the disks*

(b) *Determine the natural frequencies and natural modes of the system if $I_1 = I_2 = I$, $G_{J1} = G_{J2} = G_J$, and $l_1 = l_2 = l$*

(c) *Obtain the response of the system to the torques $M_1(t) = 0$, and $M_2(t) = M_2 e^{-\alpha t}$ in discrete time*

(d) *Obtain the response of the system to the torques $M_1(t) = 0$, and $M_2(t) = M_2 e^{-\alpha t}$*

(e) *Obtain in discrete time the response of the system to the torques $M_1(t) = 0$, and $M_2(t) = M_2 e^{-\alpha t}$ using MATLAB.*

Solution:

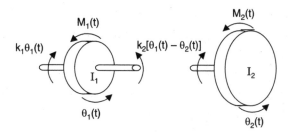

Fig. E3.9(a)

(a) The equations of motion are given by

$$I_1\ddot{\theta}_1 = M_1 - k_1\theta_1 + k_2(\theta_2 - \theta_1)$$

$$I_2\ddot{\theta}_2 = M_2 - k_2(\theta_2 - \theta_1) \tag{1}$$

where $k_i = \dfrac{GJ_i}{L_i}$, $i = 1, 2$.

Rearranging Eq. (1), we get

$$I_1\ddot{\theta}_1 + \left(\frac{GJ_1}{L_1} + \frac{GJ_2}{L_2}\right)\theta_1 - \frac{GJ_2}{L_2}\theta_2 = M_1$$

$$I_2\ddot{\theta}_2 - \frac{GJ_2}{L_2}\theta_1 + \frac{GJ_2}{L_2}\theta_2 = M_2 \tag{2}$$

In matrix form, we can write

$$\begin{bmatrix} I_1 & 0 \\ 0 & I_2 \end{bmatrix}\begin{bmatrix} \ddot{\theta}_1 \\ \ddot{\theta}_2 \end{bmatrix} + \begin{bmatrix} \dfrac{GJ_1}{L_1} + \dfrac{GJ_2}{L_2} & -\dfrac{GJ_2}{L_2} \\ -\dfrac{GJ_2}{L_2} & \dfrac{GJ_2}{L_2} \end{bmatrix}\begin{bmatrix} \theta_1 \\ \theta_2 \end{bmatrix} = \begin{bmatrix} M_1 \\ M_2 \end{bmatrix} \tag{3}$$

(b) Denoting

$$GJ_1 = GJ_2 = GJ, I_1 = I_2 = I, L_1 = L_2 = L \tag{4}$$

the equations of motion of the system [Eq. (3)] can be written as

$$M\ddot{\theta}(t) + K\theta(t) = 0 \tag{5}$$

where $M = I\begin{bmatrix} 1 & 0 \\ 0 & 1 \end{bmatrix}$, $K = \dfrac{GJ}{L}\begin{bmatrix} 2 & -1 \\ -1 & 1 \end{bmatrix}$, $\theta(t) = \begin{bmatrix} \theta_1(t) \\ \theta_1(t) \end{bmatrix}$ $\tag{6}$

are the mass matrix, stiffness matrix and configuration vector, respectively. The free vibration solution can be written as

$$\theta_i(t) = \Theta_i e^{i\omega t}, i = 1, 2 \tag{7}$$

where ω is the frequency of oscillation and $\Theta = [\Theta_1 \ \Theta_2]^T$ is a vector of constants, we have

$$\begin{bmatrix} 2 & -1 \\ -1 & 1 \end{bmatrix}\begin{bmatrix} \Theta_1 \\ \Theta_2 \end{bmatrix} = \lambda\begin{bmatrix} \Theta_1 \\ \Theta_2 \end{bmatrix}, \lambda = \omega^2\frac{IL}{GJ} \tag{8}$$

The characteristic equation can be written as

$$\begin{vmatrix} 2-\lambda & -1 \\ -1 & 1-\lambda \end{vmatrix} = \lambda^2 - 3\lambda + 1 = 0 \tag{9}$$

The eigenvalues are given by

$$\begin{matrix} \lambda_1 \\ \lambda_2 \end{matrix} = \frac{3}{2} \mp \frac{\sqrt{5}}{2} \tag{10}$$

The natural frequencies are given by

$$\omega_1 = 0.6180 \sqrt{GJ/IL} \ , \ \omega_2 = 1.6180 \sqrt{GJ/IL} \tag{11}$$

Denote the modal vector corresponding to λ_1 by $\Theta_1 = [\Theta_{11} \ \Theta_{21}]^T$, the modal vector is from the matrix equation as

$$\begin{bmatrix} 2 & -1 \\ -1 & 1 \end{bmatrix} \begin{bmatrix} \Theta_{11} \\ \Theta_{21} \end{bmatrix} = \lambda_1 \begin{bmatrix} \Theta_{11} \\ \Theta_{21} \end{bmatrix} = \frac{3-\sqrt{5}}{2} \begin{bmatrix} \Theta_{11} \\ \Theta_{21} \end{bmatrix} \tag{12}$$

or

$$\Theta_1 = \Theta_{11} \begin{bmatrix} 1 \\ 1.6180 \end{bmatrix} \tag{13}$$

In a similar way by letting $\Theta_2 = [\Theta_{12} \ \Theta_{22}]^T$, we have

$$\begin{bmatrix} 2 & -1 \\ -1 & 1 \end{bmatrix} \begin{bmatrix} \Theta_{12} \\ \Theta_{22} \end{bmatrix} = \lambda_2 \begin{bmatrix} \Theta_{12} \\ \Theta_{22} \end{bmatrix} = \frac{3+\sqrt{5}}{2} \begin{bmatrix} \Theta_{12} \\ \Theta_{22} \end{bmatrix} \tag{14}$$

or

$$\Theta_2 = \Theta_{12} \begin{bmatrix} 1 \\ 1.6180 \end{bmatrix} \tag{15}$$

The modal vectors are shown below

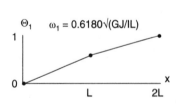

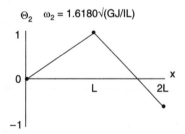

Fig. E3.9(b)   Fig. E3.9(c)

(c) The equations of motion are

$$M \ddot{\theta}(t) + K\theta(t) = M(t) \tag{16}$$

where $M = I \begin{bmatrix} 1 & 0 \\ 0 & 1 \end{bmatrix}$, $K = \dfrac{GJ}{L} \begin{bmatrix} 2 & -1 \\ -1 & 1 \end{bmatrix}$, $\theta(t) = \begin{bmatrix} \theta_1(t) \\ \theta_2(t) \end{bmatrix}$, $M(t) = \begin{bmatrix} M_1(t) \\ M_2(t) \end{bmatrix}$ (17)

The solution $\theta(t)$ is given by

$$\theta(t) = \eta_1(t) \Theta_1 + \eta_2(t) \Theta_2 \tag{18}$$

where $\eta_1(t)$ and $\eta_2(t)$ are modal coordinates and Θ_1 and Θ_2 are modal vectors. The modal equations can be written as

$$m'_{11} \ddot{\eta}_1 (t) + m'_{11}\omega_1^2 \, \eta_1(t) = N_1(t), \; m'_{22} \ddot{\eta}_2(t) + m'_{22} \, \omega_2^2 \, \eta_2(t) = N_2(t) \tag{19}$$

where
$$\omega_1 = \sqrt{\frac{3-\sqrt{5}}{2} \frac{GJ}{IL}} \; , \; \omega_2 = \sqrt{\frac{3+\sqrt{5}}{2} \frac{GJ}{IL}} \tag{20}$$

The natural frequencies are given by Eq. (20).

$$\Theta_1 = \begin{bmatrix} 1 \\ \dfrac{1+\sqrt{5}}{2} \end{bmatrix}, \quad \Theta_2 = \begin{bmatrix} 1 \\ \dfrac{1-\sqrt{5}}{2} \end{bmatrix} \tag{21}$$

The modal vectors are given by Eq. (21).

$$m'_{11} = \Theta_1^T M \, \Theta_1 = \begin{bmatrix} 1 \\ \dfrac{1+\sqrt{5}}{2} \end{bmatrix}^T \begin{bmatrix} I & 0 \\ 0 & I \end{bmatrix} \begin{bmatrix} 1 \\ \dfrac{1+\sqrt{5}}{2} \end{bmatrix} = \frac{5+\sqrt{5}}{2} \, I$$

$$m'_{22} = \Theta_2^T M \, \Theta_2 = \begin{bmatrix} 1 \\ \dfrac{1-\sqrt{5}}{2} \end{bmatrix}^T \begin{bmatrix} I & 0 \\ 0 & I \end{bmatrix} \begin{bmatrix} 1 \\ \dfrac{1-\sqrt{5}}{2} \end{bmatrix} = \frac{5-\sqrt{5}}{2} \, I \tag{22}$$

The modal mass coefficients are given by Eq.(22).

$$N_1(t) = \Theta_1^T \, M(t) = \begin{bmatrix} 1 \\ \dfrac{1+\sqrt{5}}{2} \end{bmatrix}^T \begin{bmatrix} 0 \\ M_2 e^{-\alpha t} \end{bmatrix} = \frac{1+\sqrt{5}}{2} \, M_2 e^{-\alpha t}$$

$$N_2(t) = \Theta_2^T \, M(t) = \begin{bmatrix} 1 \\ \dfrac{1-\sqrt{5}}{2} \end{bmatrix}^T \begin{bmatrix} 0 \\ M_2 e^{-\alpha t} \end{bmatrix} = \frac{1-\sqrt{5}}{2} \, M_2 e^{-\alpha t} \tag{23}$$

The modal forces are given by Eq. (23). The solutions $\eta_1(t)$ and $\eta_2(t)$ of the modal equations are written in the form of the convolution integrals

$$\eta_1(t) = \frac{1}{m'_{11}\omega_1} \int_o^t N_1 \, (t-\tau) \sin \omega_1 \tau \, d\tau$$

$$= \frac{1+\sqrt{5}}{(5+\sqrt{5})(\alpha^2+\omega_1^2)} \frac{M_2}{I} \left[e^{-\alpha t} - \left(\cos\omega_1 t - \frac{\alpha}{\omega_1} \sin\omega_1 t \right) \right]$$

$$\eta_2(t) = \frac{1-\sqrt{5}}{(5-\sqrt{5})(\alpha^2+\omega_2^2)} \frac{M_2}{I} \left[e^{-\alpha t} - \left(\cos\omega_2 t - \frac{\alpha}{\omega_2} \sin\omega_2 t \right) \right] \tag{24}$$

where ω_1 and ω_2 are given above. Hence, the response can be written as

$$\theta_1(t) = \eta_1(t) + \eta_2(t), \quad \theta_2(t) = \frac{1+\sqrt{5}}{2} \, \eta_1(t) + \frac{1-\sqrt{5}}{2} \, \eta_2(t) \tag{25}$$

(d) The equations of motion are

$$M\ddot{\theta} \, (t) + K\theta(t) = M(t) \tag{26}$$

where $M = I \begin{bmatrix} 1 & 0 \\ 0 & 1 \end{bmatrix}$, $K = \dfrac{GJ}{L} \begin{bmatrix} 2 & -1 \\ -1 & 1 \end{bmatrix}$, $\theta(t) = \begin{bmatrix} \theta_1(t) \\ \theta_2(t) \end{bmatrix}$, $M(t) = \begin{bmatrix} 0 \\ M_2 e^{-\alpha t} \end{bmatrix}$ $\tag{27}$

Assuming a solution of the form

$$\theta(t) = \eta_1(t) \, \Theta_1 + \eta_2(t) \, \Theta_2 \tag{28}$$

in which $\eta_1(t)$ and $\eta_2(t)$ are modal coordinates and

$$\Theta_1 = \begin{bmatrix} 1 \\ \dfrac{1+\sqrt{5}}{2} \end{bmatrix}, \quad \Theta_2 = \begin{bmatrix} 1 \\ \dfrac{1-\sqrt{5}}{2} \end{bmatrix} \tag{29}$$

are the modal vectors. The modal equations are given by

$$m'_{11}\ddot{\eta}_1(t) + m'_{11}\,\omega_1^2\,\eta_1(t) = N_1(t)$$

$$m'_{22}\ddot{\eta}_2(t) + m'_{22}\,\omega_2^2\,\eta_2(t) = N_2(t) \tag{30}$$

in which $\omega_1 = \sqrt{\dfrac{3-\sqrt{5}}{2}\dfrac{GJ}{IL}}$, $\omega_2 = \sqrt{\dfrac{3+\sqrt{5}}{2}\dfrac{GJ}{IL}}$ $\tag{31}$

are the natural frequencies.

$$m'_{11} = \frac{1+\sqrt{5}}{2}\,I, \quad m'_{22} = \frac{1-\sqrt{5}}{2}\,I \tag{32}$$

are modal mass coefficients.

The modal forces are given by

$$N_1(t) = \frac{1+\sqrt{5}}{2}\,M_2 e^{-\alpha t}, \quad N_2(t) = \frac{1-\sqrt{5}}{2}\,M_2 e^{-\alpha t} \tag{33}$$

The response is given by

$$\theta(n) = \eta_1(n)\Theta_1 + \eta_2(n)\Theta_2, \quad n = 1, 2, \ldots \tag{34}$$

where $\eta_1(n) = \displaystyle\sum_{k=0}^{n} N_1(k)\,g_1(n-k),$

$$\eta_2(n) = \sum_{k=0}^{n} N_2(k)\,g_2(n-k), \quad n = 1, 2, \ldots \tag{35}$$

are the discrete-time modal coordinates given in the form of convolution sums, in which the discrete time impulse responses are given by

$$g_i(n) = \frac{T}{m'_{ii}\omega_i}\,\sin n\omega_i T, \quad i = 1, 2 \tag{36}$$

where T is the sampling period. The discrete-time response is given by

$$\theta(n) = T \sum_{k=0}^{n} \left\{ \left[\frac{N_1(k)}{m'_{11}\,\omega_1}\sin(n-k)\,\omega_1 T \right]\Theta_1 + \left[\frac{N_2(k)}{m'_{22}\,\omega_2}\sin(n-k)\,\omega_2 T \right]\Theta_2 \right\}$$

$$= \frac{TM_2}{1\sqrt{GJ/IL}} \sum_{k=0}^{n} \left\{ e^{-\alpha kT}\sin(n-k)\,0.618034\sqrt{\frac{GJ}{IL}}\,T \begin{bmatrix} 0.7236907 \\ 1.17082 \end{bmatrix} \right.$$

$$\left. + e^{-\alpha kT}\sin(n-k)\,1.618034\sqrt{GJ/IL}\,T \begin{bmatrix} -0.276393 \\ 0.17082 \end{bmatrix} \right\} \tag{37}$$

Denoting $\sqrt{GJ/IL} = 1$, $M_2/I = 1$, $\alpha = 1$ and $T = 0.01$ s, the response is given by

$$\theta(n) = 0.01 \sum_{k=0}^{n} e^{-0.01k} \left\{ \sin 0.618034\,(n-k) \begin{bmatrix} 0.723607 \\ 1.170820 \end{bmatrix} \right.$$

$$\left. + \sin 1.618034(n-k)T \begin{bmatrix} -0.276393 \\ 0.170820 \end{bmatrix} \right\} \tag{38}$$

The discrete-time response sequence is given by

$$\theta(0) = \begin{bmatrix} 0 \\ 0 \end{bmatrix}$$

$$\theta(1) = 0.01 \left\{ \sin 0.00618034 \begin{bmatrix} 0.723607 \\ 1.170820 \end{bmatrix} + \sin 0.0161803 \begin{bmatrix} -0.276393 \\ 1.170820 \end{bmatrix} \right\}$$

$$= \begin{bmatrix} 1.82291 \times 10^{-9} \\ 9.999 \times 10^{-5} \end{bmatrix}$$

$$\theta(2) = 0.01 \left\{ [\sin(0.00618034 \times 2) + e^{-0.01} \sin 0.00618034] \begin{bmatrix} 0.723607 \\ 1.170820 \end{bmatrix} \right.$$

$$\left. + [\sin(0.0161803 \times 2) + e^{-0.01} \sin 0.0161803] \begin{bmatrix} -0.276393 \\ 0.170820 \end{bmatrix} \right\}$$

$$= \begin{bmatrix} 1.54497 \times 10^{-8} \\ 2.990 \times 10^{-4} \end{bmatrix} \tag{39}$$

(e) The response $\theta_i(n)$ ($i = 1, 2$) is plotted in Fig. E3.9 (d) obtained from the following MATLAB program.

```
% Response of 2-degree of freedom system
clear
clf
I=1; % mass
k=1;%=GJ/L torsional stiffness
M=I*[1 0;0 1];% mass matrix
K=k*[2 -1;-1 1];%stiffness matrix
[u,W]=eig(K,M);% eigenvalue problem
% W= eigenvalues
u(:,1)=u(:,1)/max(u(:,1)); % normalization
u(:,2)=u(:,2)/max(u(:,2));
[w(1),I1]=min(max(W)); % relabeling of the eigenvalues
[w(2),I2]=max(max(W));
w(1)=sqrt(w(1)); % Lowest natural frequency
w(2)=sqrt(w(2)); % highest natural frequency
U(:,1)=u(:,I1); % relabelling of the eigenvectors
U(:,2)=u(:,I2);
m1=U(:,1)'*M*U(:,1); % mass quantities
m2=U(:,2)'*M*U(:,2);
T=0.01; % sampling period
```

```
N=2000; % sampling times
M2=1; % second disk torque amplitude
alpha=1;
n=[1:N];
N1=U(:,1)'*[zeros(1,N);M2*exp(-alpha*n*T)]; % modal forces
N2=U(:,2)'*[zeros(1,N);M2*exp(-alpha*n*T)];
g1=T*sin((n-1)*w(1)*T)/(m1*w(1)); %discrete time impulse responses
g2=T*sin((n-1)*w(2)*T)/(m2*w(2));
c1=conv(N1,g1); %convolution sum
c2=conv(N2,g2);
theta=U(:,1)*c1(1:N)+U(:,2)*c2(1:N); % N samples for plotting
n=[0:N-1];
axes('position',[0.1 0.2 0.8 0.7])
plot(n,theta(1,:),'.',n,theta(2,:),'.')
h=title('Response by the convolution sum');
set(h,'FontName',  'Times','FontSize',12)
h=xlabel('n')
set(h,'FontName','Times','FontSize',12)
h=ylabel('\theta_1(n),\theta_2(n)');
set(h,'FontName','Times','FontSize',12)
grid
```

Its output is shown in Fig. E3.9(d).

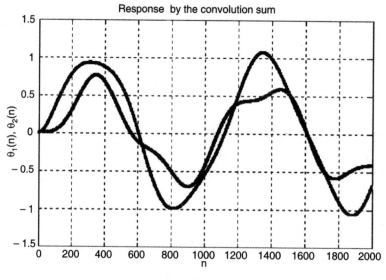

Fig. E3.9(d)

Example 3.10. *Obtain the response of the system of Problem E3.9 to the initial excitation* $\theta_1(0) = 0$, $\theta_2(0) = 1.5$, $\dot{\theta}_1(0) = 1.8 \sqrt{GJ / Il}$, *and* $\dot{\theta}_2 (0) = 0$. *Plot the response of the system using MATLAB.*

Solution:

The initial conditions are given as

$$\theta_1(0) = 0, \ \theta_2(0) = 1.5, \ \dot{\theta}_1(0) = 1.8\sqrt{\frac{GJ}{IL}}, \ \dot{\theta}_2(0) = 0 \tag{1}$$

From Problem V3.9, we have

$$\omega_1 = 0.6180\sqrt{\frac{GJ}{IL}}, \ \Theta_1 = \begin{bmatrix} \Theta_{11} \\ \Theta_{21} \end{bmatrix} = \Theta_{11}\begin{bmatrix} 1 \\ 1.6180 \end{bmatrix}$$

$$\omega_2 = 1.6180\sqrt{\frac{GJ}{IL}}, \ \Theta_2 = \begin{bmatrix} \Theta_{12} \\ \Theta_{22} \end{bmatrix} = \Theta_{12}\begin{bmatrix} 1 \\ 1.6180 \end{bmatrix} \tag{2}$$

The response to the initial excitation is a superposition of the natural modes. Hence

$$\theta(t) = C_1 \cos(\omega_1 t - \phi_1)\Theta_1 + C_2 \cos(\omega_2 t - \phi_2)\Theta_2$$

or

$$\theta(t) = C_1 (\cos \omega_1 t \cos \phi_1 + \sin \omega_1 t \sin \phi_1)\Theta_1$$
$$+ C_2(\cos \omega_2 t \cos \phi_2 + \sin \omega_2 t \sin \phi_2)\Theta_2 \tag{3}$$

and

$$\dot{\theta}(t) = C_1\omega_1(\sin \omega_1 t \cos \phi_1 - \cos \omega_1 t \sin \phi_1)\Theta_1$$
$$- C_2\omega_2(\sin \omega_2 t \cos \phi_2 - \cos \omega_2 t \sin \phi_2)\Theta_2$$

If $t = 0$, then Eq. (3) becomes

$$\theta(0) = \begin{bmatrix} \theta_1(0) \\ \theta_2(0) \end{bmatrix} = C_1 \cos \phi_1 \begin{bmatrix} \Theta_{11} \\ \Theta_{21} \end{bmatrix} + C_2 \cos \phi_2 \begin{bmatrix} \Theta_{12} \\ \Theta_{22} \end{bmatrix}$$

$$\dot{\theta}(0) = \begin{bmatrix} \dot{\theta}_1(0) \\ \dot{\theta}_2(0) \end{bmatrix} = C_1\omega_1 \sin \phi_1 \begin{bmatrix} \Theta_{11} \\ \Theta_{21} \end{bmatrix} + C_2\omega_2 \sin \phi_2 \begin{bmatrix} \Theta_{12} \\ \Theta_{22} \end{bmatrix} \tag{4}$$

From Eq. (4) we have,

$$C_1 \cos \phi_1 = \frac{\Theta_{22}\theta_{10} - \Theta_{12}\theta_{20}}{|\Theta|}, \ C_2 \cos \phi_2 = \frac{\Theta_{11}\theta_{20} - \Theta_{21}\theta_{10}}{|\Theta|}$$

$$C_1 \sin \phi_1 = \frac{\Theta_{22}\dot{\theta}_{10} - \Theta_{12}\dot{\theta}_{20}}{\omega_1 |\Theta|}, \ C_2 \sin \phi_2 = \frac{\Theta_{11}\dot{\theta}_{20} - \Theta_{21}\dot{\theta}_{10}}{\omega_2 |\Theta|} \tag{5}$$

where $\theta_{10} = \theta_1(0)$, $\theta_{20} = \theta_2(0)$, $\dot{\theta}_{10} = \dot{\theta}_1(0)$, $\dot{\theta}_{20} = \dot{\theta}_2(0)$ and $|\Theta|$ is the determinant of the matrix

$$\Theta = \begin{bmatrix} \Theta_{11} & \Theta_{12} \\ \Theta_{21} & \Theta_{22} \end{bmatrix} \tag{6}$$

Letting $\Theta_{11} = \Theta_{12} = 1$,

$$C_1 \cos \phi_1 = \frac{1.5}{|\Theta|}, \ C_2 \cos \phi_2 = \frac{1.5}{|\Theta|},$$

$$C_1 \sin \phi_1 = \frac{1.8}{|\Theta|}, \ C_2 \sin \phi_2 = \frac{1.8}{|\Theta|}$$

$$|\Theta| = \begin{vmatrix} 1 & 1 \\ 1.6180 & -1.6180 \end{vmatrix} = -2.2360 \tag{7}$$

The response to the given initial excitation is given by

$$\theta(t) = \begin{bmatrix} \theta_1(t) \\ \theta_2(t) \end{bmatrix} = \frac{1}{2.2360}\left[(1.5\cos\omega_1 t + 1.8\cos\omega_1 t)\begin{bmatrix} 1 \\ 1.6180 \end{bmatrix} \right.$$

$$\left. + (-1.5\cos\omega_2 t + 1.8\sin\omega_2 t)\begin{bmatrix} 1 \\ -0.6180 \end{bmatrix} \right]$$

or by components

$$\theta_1(t) = 0.6708\left(\cos 0.6180\sqrt{\frac{GJ}{IL}}\,t - \cos 1.6180\sqrt{\frac{GJ}{IL}}\,t \right)$$

$$+ 0.8060\left(\sin 0.6180\sqrt{\frac{GJ}{IL}}\,t + \sin 1.6180\sqrt{\frac{GJ}{IL}}\,t \right)$$

$$\theta_2(t) = 1.0854\cos 0.6180\sqrt{\frac{GJ}{IL}}\,t + 1.3025\sin 0.6180\sqrt{\frac{GJ}{IL}}\,t$$

$$+ 0.4146\cos 1.6180\sqrt{\frac{GJ}{IL}}\,t - 0.4975\sin 106180\sqrt{\frac{GJ}{IL}}\,t \tag{8}$$

The MATLAB program listed as follows:

```
% response of a two-degree-of freedom system to initial excitations
clear
clf
I=1; % inertia
k=1;%=GJ/L stiffness
M=I*[1 0;0 1];% mass
K=k*[2 -1;-1 1];%stiffness
[u,W]=eig(K,M);% eigenvalue problem
% W=matrix of eigenvalues
u(:,1)=u(:,1)/max(u(:,1)); % normalization
u(:,2)=u(:,2)/max(u(:,2));
[w(1),I1]=min(max(W)); % relabeling
[w(2),I2]=max(max(W));
w(1)=sqrt(w(1)); % lowest natural frequency
w(2)=sqrt(w(2)); % highest natural frequency
U(:,1)=u(:,I1); % relabelling
U(:,2)=u(:,I2);
x0=[0;2];% Initial displacement
v0=[2*sqrt(k/I);0]; % initial velocity
t=[0:0.1:20]; % initial time, time increment, final time
% displacement
x1=(((U(2,2)*x0(1)-U(1,2)*x0(2))*cos(w(1)*t)+(U(2,2)*v0(1)-
U(1,2)*v0(2))*sin(w(1)*t)/w(1))*U(1,1)+((U(1,1)*x0(2)
-U(2,1)*x0(1))*cos(w(2)*t)+(U(1,1)*v0(2)-U(2,1)*v0(1))*sin(w(2)*t)
/w(2))*U(1,2))/det(U);
x2=(((U(2,2)*x0(1)-U(1,2)*x0(2))*cos(w(1)*t)+(U(2,2)*v0(1)-
```

```
U(1,2)*v0(2))*sin(w(1)*t)/w(1))*U(2,1)+((U(1,1)*x0(2)-
U(2,1)*x0(1))*cos(w(2)*t)+(U(1,1)*v0(2)-U(2,1)*v0(1))*sin(w(2)*t)
/w(2))*U(2,2))/det(U);
axes('position',[0.2 0.3 0.6 0.5])
plot(t,x1,t,x2)
title('Response to initial excitation')
ylabel('\theta_1(t),\theta_2(t)')
xlabel('t[s]')
legend('\theta_1(t)','\theta_2(t)',1)
grid
```

The corresponding output obtained is shown in Fig.E 3.10

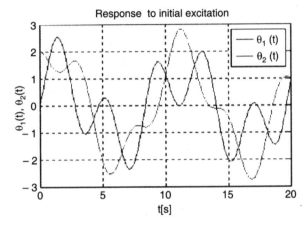

Fig. E3.10

Example 3.11. *A simplified model of an automobile suspension system is shown in Fig. E3.11 as a two degree of freedom system. Write a MATLAB script to determine the natural frequencies of this model.*

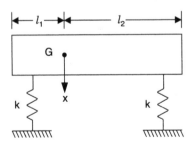

Fig. E3.11 Simplified model of an automobile.

The differential equations governing the motion of the system are given as

$$\begin{bmatrix} m & 0 \\ 0 & 1 \end{bmatrix}\begin{bmatrix} \ddot{x} \\ \ddot{\theta} \end{bmatrix} + \begin{bmatrix} 2k & (l_2 - l_1)k \\ (l_2 - l_1)k & (l_2^2 - l_1^2)k \end{bmatrix}\begin{bmatrix} x \\ \theta \end{bmatrix} = \begin{bmatrix} 0 \\ 0 \end{bmatrix}$$

where x is the displacement of the mass center and θ is the angular rotation of the body from its horizontal position.

The parameters are given as

Automobile weight, W = 5000 lb

Centroidal moment of inertia, I = 400 slug-ft^2

Spring stiffness, k = 2500 lb/ft

$$l_1 = 3.4 \text{ ft}$$

$$l_2 = 4.6 \text{ ft}$$

Solution:

The MATLAB program is given as follows:

```
%Two-degree-of-freedom system
W=input('Vehicle weight in lb');
I=input('Mass moment of inertia in slugs-ft^2')
k=input('Stiffness in lb/ft')
a=input('Distance from rear springs to cg in ft')
b=input('Distance from front springs to cg')
% mass matrix
g=32.2;
m=W/g;
M=[m,0;0,I];
% stiffness matrix
K=[2*k,(b-a)*k;(b-a)*k,(b^2+a^2)*k];
% eigenvalues and eigenvectors calculation
C=inv(M)*K;
[V,D]=eig(C);
om_1=sqrt(D(1,1));
om_2=sqrt(D(2,2));
X1=[V(1,1);V(2,1)];
X2=[V(1,2);V(2,2)];
% Output
disp('Vehicle weight in lb='); disp(W)
disp('moment of inertia in slugs-ft^2');disp(I)
disp('Stiffness in lb/ft='); disp(k)
disp('Distance from rear springs to cg in ft='); disp(a)
disp('Distance from front springs to cg in ft=');disp(b)
disp('Mass-matrix');disp(M)
disp('Stiffness-matrix');disp(K)
disp('Natural frequencies in rad/s=');
disp(om_1)
disp(om_2)
disp('Mode shape vectors'); disp(X1)
disp(X2)
```

The output of this program is as follows:

```
Vehicle weight in lb   5000
W=5000
Mass moment of inertia in slugs-ft^2      400
```

```
I =  400
Stiffness in lb/ft 2500
k = 2500
Distance from rear springs to cg gravity in ft 3.4
a = 3.4000
Distance from front springs to cg 4.6
b = 4.6000
Vehicle weight in lb= 5000
Moment of inertia in slugs-ft^2    400
Stiffness in lb/ft=2500
Distance from rear springs to cg in ft=3.4000
Distance from front springs to cg in ft=4.6000
Mass-matrix
      155.2795             0
         0          400.0000
Stiffness-matrix
   1.0e+004 *
      0.5000    0.3000
      0.3000    8.1800
Natural frequencies in rad/s=
      5.6003
     14.3296
Mode shape vectors
    -0.9991
     0.0433
    -0.1109
    -0.9938
```

Example 3.12. *Determine the free-vibration response of a two degree of freedom system shown in Fig. E3.12 with the initial conditions $x_1(0) = 0$, $x_2(0) = 0.005$ m, $\dot{x}_1(0) = 0$, $\dot{x}_2(0) = 0$. The parameters of the system are given as $m = 30$ kg, $k = 20,000$ N/m, and $c = 150$ N.s/m.*

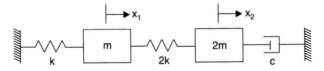

Fig. E3.12 Two degree of freedom system.

The differential equations governing the motion of the system are

$$\begin{bmatrix} m & 0 \\ 0 & 2m \end{bmatrix}\begin{bmatrix} \ddot{x}_1 \\ \ddot{x}_2 \end{bmatrix} + \begin{bmatrix} 0 & 0 \\ 0 & c \end{bmatrix}\begin{bmatrix} \dot{x}_1 \\ \dot{x}_2 \end{bmatrix} + \begin{bmatrix} 3k & -2k \\ -2k & 2k \end{bmatrix}\begin{bmatrix} x_1 \\ x_2 \end{bmatrix} = \begin{bmatrix} 0 \\ 0 \end{bmatrix}$$

or $\qquad M\dot{y} + Ky = 0$

where $\qquad M = \begin{bmatrix} 0 & M \\ M & c \end{bmatrix}; K = \begin{bmatrix} -M & 0 \\ 0 & K \end{bmatrix}; Y = \begin{bmatrix} \dot{x} \\ x \end{bmatrix}$

The solution is assumed as

$$y = \phi e^{-\gamma t}$$

where γ are the eigenvalues of $M^{-1}K$ and ϕ are the eigenvectors. The general solution is a linear combination over all solutions, that is,

$$y = \sum_{j=1}^{4} c_j \phi_j e^{-\gamma_j t}$$

and application of initial conditions gives

$$y_0 = \sum_{j=1}^{4} c_j \phi_j = VC$$

and

$$C = V^{-1} y_0$$

Solution:

The MATLAB program is given as follows:

```
m=30; % Mass
k=20000; % Stiffness
c=150; % Damping
% 4 x 4 matrices
disp('4 x 4 Mass matrix');
mt=[0,0,m,0;0,0,0,2*m;m,0,0,0;0,2*m,0,c];
disp('4 x 4 stiffness matrix');
kt=[-m,0,0,0;0,-2*m,0,0;0,0,3*k,-2*k;0,0,-2*k,2*k];
Z=inv(mt)*kt;
[V,D]=eig(Z);
disp('Eigenvalues');
V
disp('Initial conditions');
x0=[0;0;0.005;0]
disp('Integration constants');
S=inv(V)*x0
tk=linspace(0,2,101);
% Evaluation of time dependent response
% Recall that x1=y3 and x2=y4
for k=1:101
  t=tk(k);
  for i=3:4
    x(k,i-2)=0;
    for j=1:4
     x(k,i-2)=x(k,i-2)+(real(S(j))*real(V(i,j))-imag(S(j))*imag(V(i,j)))
     *cos(imag(D(j,j))*t);
     x(k,i-2)=x(k,i-2)+(imag(S(j))*real(V(i,j))-real(S(j))
     *imag(V(i,j)))*sin(imag(V(i,j))*t);
    x(k,i-2)=x(k,i-2)*exp(-real(D(j,j))*t);
    end
```

```
        end
end
plot(tk,x(:,1),'-',tk,x(:,2),':')
title('Solution of problem E3.12')
xlabel('t[sec]')
ylabel('x(m)')
legend('x1(t)','x2(t)')
```

The output of this program is given below. See also Fig.E3.12(a).

```
V =

  -0.9390            -0.9390            0.5886 - 0.0085i   0.5886 + 0.0085i
   0.3428 - 0.0185i   0.3428 + 0.0185i  0.8050             0.8050
   0.0001 - 0.0188i   0.0001 + 0.0188i -0.0026 + 0.0440i  -0.0026 - 0.0440i
   0.0003 + 0.0069i   0.0003 - 0.0069i -0.0044 + 0.0601i  -0.0044 - 0.0601i
```

Initial conditions

```
x0 =
        0
        0
   0.0050
        0
```

Integration constants

```
S =
  -0.0013 + 0.1048i
  -0.0013 - 0.1048i
  -0.0019 - 0.0119i
  -0.0019 + 0.0119i
```

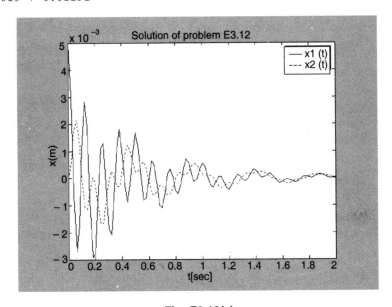

Fig. E3.12(a)

Example 3.13. *For systems with arbitrary viscous-damping, the response must be obtained in the state-space, which implies the use of the transition-matrix. If the response is to be evaluated on a computer, then the state-equations must be transformed to discrete time. Determine the free-vibration response of a 2-degree-of-freedom damped system with initial conditions X(0) = {0, 0.01} and $\dot{X}$(0) = {0, 0}. Given*

$$[M] = \begin{bmatrix} 0 & 0 & 30 & 0 \\ 0 & 0 & 0 & 50 \\ 30 & 0 & 0 & 0 \\ 0 & 50 & 0 & 80 \end{bmatrix}$$

$$[K] = \begin{bmatrix} -40 & 0 & 0 & 0 \\ 0 & -50 & 0 & 0 \\ 0 & 0 & 35000 & -25000 \\ 0 & 0 & -25000 & 4000 \end{bmatrix}$$

Solution: The solution is similar to the problem E3.12, and the MATLAB program is written as follows:

```
mt=[0  0  30  0;0,0,0,50;30,0,0,0;0,50,0,80];
kt=[-40,0,0,0;0,-50,0,0;0,0,35000,-25000;0,0,-25000,  4000];
Z=inv(mt)*kt;
[V,D]=eig(Z);
disp('Eigenvalues')
DS=[D(1,1),D(2,2),D(3,3),D(4,4)]
disp('Eigenvectors')
V
x0=[0;0;0.01;0];
S=inv(V)*x0;
tk=linspace(0,2,101);
for k=1:101
  t=tk(k);
  for i=3:4
  x(k,i-2)=0;
  for j=1:4
x(k,i-2)=x(k,i-2)+(real(S(j))*real(V(i,j))
-imag(S(j))*imag(V(i,j)))*cos(imag(D(j,j))*t);
x(k,i-2)=x(k,i-2)+(imag(S(j))*real(V(i,j))-imag(S(j))*imag(V(i,j)))
*sin(imag(V(i,j))*t);
x(k,i-2)=x(k,i-2)*exp(-real(D(j,j))*t);
end
end
end
plot(tk,x(:,1),'-',tk,x(:,2),':')
title('Free Vibration response of damped system')
xlabel('t (sec)')
```

```
ylabel('x (m)')
legend('x1(t)','x2(t)')
```

The output obtained is given as follows:

```
Eigenvalues
DS =
```

Columns 1 through 2

```
   1.1082e-001 +4.4615e+001i   1.1082e-001 -4.4615e+001i
```

Columns 3 through 4

```
  -1.5162e+001                 1.6541e+001
Eigenvectors
V =
Columns 1 through 2
   9.5465e-001                 9.5465e-001
  -2.9638e-001 +9.4404e-003i  -2.9638e-001 -9.4404e-003i
  -6.3783e-005 +2.5677e-002i  -6.3783e-005 -2.5677e-002i
  -1.9510e-004 -6.6437e-003i  -1.9510e-004 +6.6437e-003i
Columns 3 through 4
   4.6275e-001                -4.5475e-001
   8.8381e-001                -8.8839e-001
   3.6624e-002                 3.2991e-002
   5.8290e-002                 5.3710e-002
```

Fig. E3.13 shows the response in time-domain obtained from the output of the MATLAB program.

Note: Here $M = \begin{bmatrix} 25 & 0 \\ 0 & 50 \end{bmatrix}$, $C = \begin{bmatrix} 0 & 0 \\ 0 & 80 \end{bmatrix}$ and

$$K = \begin{bmatrix} 35000 & -25000 \\ -25000 & 3000 \end{bmatrix}$$ and state matrices are

respectively $M^T = \begin{bmatrix} 0 & 0 & 25 & 0 \\ 0 & 0 & 0 & 50 \\ 25 & 0 & 0 & 0 \\ 0 & 50 & 0 & 80 \end{bmatrix}$

and $K^T = \begin{bmatrix} -30 & 0 & 0 & 0 \\ 0 & -50 & 0 & 0 \\ 0 & 0 & 35000 & -25000 \\ 0 & 0 & -25000 & 3000 \end{bmatrix}$

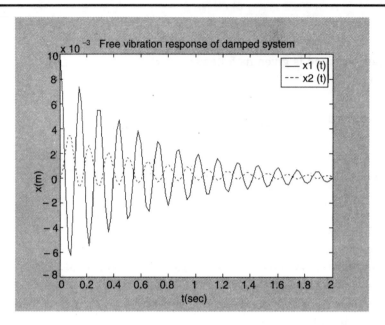

Fig. E3.13

Example 3.14. *In the Exampie E3.13, if a force $F_o exp(-\alpha t)$ acts on the system, find the forced vibration response using the MATLAB program. Given $F_0 = 60$.*

Solution:

Here the first few steps are common as in free-vibration response problem.

```
syms t tau
m=25;
k=12500;
c=80;
F0=60;
alpha=1.5;
mt=[0,0,m,0;0,0,0,2*m;m,0,0,0;0,2*m,0,c];
kt=[-m,0,0,0;0,-2*m,0,0;0,0,3*k,-2*k;0,0,-2*k,2*k];
z=inv(mt)*kt;
[V,D]=eig(z);
L=conj(V)'*mt*V;
for j=1:4
  ss=1/sqrt(L(j,j));
  for i=1:4
  P(i,j)=V(i,j)*ss;
  end
end
F=[0;0;0;F0*exp(-alpha*tau)];
G=P'*F;
G=vpa(G);
```

```
%Convolution integral solution
for i=1:4
  f(i)=G(i)*exp(-D(i,i)*(t-tau));
  p(i)=int(f(i),tau,0,t);
end
disp('solution for modal coordinates')
p=[p(1);p(2);p(3);p(4)];  disp(p)
y=P*p;
disp('response')
disp('x1=y3, x2=y4 ')
y=vpa(y);
% Plotting the system response
time=linspace(0,1.5,101);
for k=1:101
  x1a=subs(y(3),t,time(k));
  x2a=subs(y(4),t,time(k));
  x1b(k)=vpa(real(x1a));
  x2b(k)=vpa(real(x2a));
end
x1=double(x1b);
x2=double(x2b);
plot(time,x1,'-',time,x2,':')
xlabel('t(seconds)')
ylabel('response(m)')
legend('x1(t)',  'x2(t)')
```

The output of the program is shown as the forced vibration response in Fig. E3.14.

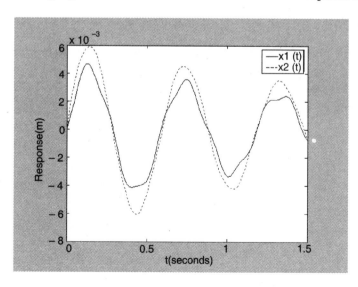

Fig. E3.14

Note: In the above program, a few additional MATLAB functions have been used. They are described as follows:

syms t tau or *syms('t', 'tau')* defines the symbolic variables.

Variable substitution in symbolic expressions are performed with the function *'subs'*.

subs(f,x,s) replaces x by s in the expression f.

'int' function integrates a symbolic expression or the elements of a symbolic array.

int(f,s,a,b) finds symbolic expressions for the definite integral from a to b with respect to symbolic variable s.

'vpa' function evaluates a single symbolic expression or character string to the default or specified accuracy.

Example 3.15. *Two gears A and B in mesh are mounted on two uniform circular shafts of equal stiffness* $\dfrac{GJ}{L}$. *If the gear A is subjected to a torque $M_0 \cos \omega t$, derive an expression for angular motion of B. Assume the radius ratio as:* $\dfrac{R_A}{R_B} = n$. *Here L is length of each shaft. Write a MATLAB script to plot the response.*

Solution:

Here the equation of motion is given by

$$I_{eq}\, \ddot{\theta}_A + k_{eq}.\theta_A = M_A = M_0 \cos \omega t \tag{1}$$

where the equivalent stiffness of the gears $k_{eq} = \dfrac{GJ}{L}(1 + n^2)$ and equivalent moment of inertia of gears $I_{eq} = I_A + n^2 I_B$.

Simplifying the above equation of motion we get:

$$\ddot{\theta}_A + \omega_n^2\, \theta_A = \frac{M_0}{I_A + n^2 I_B}\cos \omega t, \text{ and } \omega_n^2 = \frac{GJ\,(1 + n^2)}{L\,(I_A + n^2 I_B)} \tag{2}$$

Since $\theta_B = n\, \theta_A$, the solution is given by

$$\theta_B = \frac{M_0 L n}{GJ\,(1 + n^2)\,[1 - (\omega/\omega_n)^2]}\cos \omega t$$

The MATLAB program to plot the values of amplitude of θ_B for various values of ω is given as follows:

```
M0=1;% amplitude of the moment
L=1; %  length of shaft
GJ=1;%  torsional stiffness
n=3;%  gear ratio
r=[0:0.01:3];% frequency ratio
thetab=(M0*L*n)./(GJ*(1+(n.^n))*(1-r.^2)); % amplitude
plot(r,thetab)
title('Response to torque')
ylabel('\theta_b')
xlabel('\omega/\omega_n')
grid
```

the output is shown in Fig. E3.15 (a).

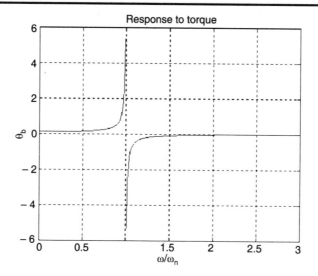

Fig. E3.15(a)

Example 3.16. *Derive the response of a viscously damped single-degree of freedom system to the trapezoidal pulse shown in Fig. E3.16. Plot response for system parameters, m = 15 kg, c = 25 NS/m and k = 5000 N/m. Use convolution sum.*

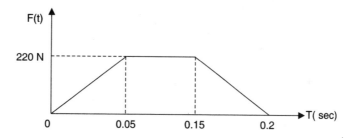

Fig. E3.16

Solution:

The system is described by:

$$\ddot{x} + 2\xi\omega_n \dot{x} + \omega_n^2 x = \frac{F(t)}{m}$$

where
$$F(t) = \begin{cases} \dfrac{2F_0}{T}t, & 0 < t < \dfrac{T}{2} \\[2mm] F_0, & \dfrac{T}{2} < t < \dfrac{3T}{2} \\[2mm] 2F_0\left(2 - \dfrac{t}{T}\right), & \dfrac{3T}{2} < t < 2T \\[2mm] 0, & t > 2T \end{cases}$$

where $T = 0.2$ sec in Fig. E3.16.

The discrete time response by convolution sum is:

$$x(n) = \sum_{k=0}^{n} F(k)\, g(n-k)$$

The MATLAB script for this problem is given below:

```
m=15; % mass
c=25; % damping
k=5000; % stiffness
F0=220;
T=0.2;
wn=sqrt(k/m); % Natural frequency
zeta=c/(2*sqrt(m*k));
Ts=0.003; % Sampling period
N=201; % sampling times
wd=wn*sqrt(1-zeta^2); % frequency
% force
for n=1:N,
if  n<=(T/2)/Ts+1;F(n)=2*F0*(n-1)*Ts/T;  else;F(n)=F0;end
if  n>(3*T/2)/Ts+1;F(n)=2*F0*(2-(n-1)*Ts/T);end
if  n>2*T/Ts+1;F(n)=0;end
end
n=[1:N];
g=Ts*exp(-(n-1)*zeta*wn*Ts).*sin((n-1)*wd*Ts)/(m*wd);
% discrete-time impulse response
c0=conv(F,g); % Convolution sum
c=c0(1:N); % plot to N samples
n=[0:N-1];
axes('position',[0.1 0.2 0.8 0.7])
plot(n,c,'.');
title('Response to the Trapezodial pulse');
xlabel('n')
ylabel('x(n) [m]')
grid
```

Output of this program is the Fig. E3.16(a)

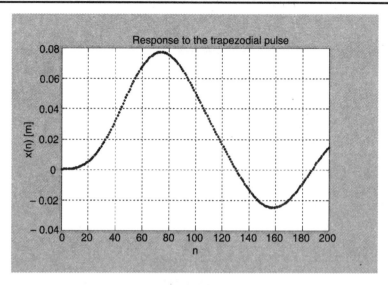

Fig. E3.16(a)

Example 3.17. *A two story building is undergoing a horizontal motion y(t)=Y_0 sin ωt. Derive expression for displacement of second floor. Write MATLAB script to plot the response. Assume appropriate values of stiffness and mass of the system.*

Equations of motion for building can be written as:

$$\begin{bmatrix} 2 & 0 \\ 0 & 2 \end{bmatrix} \begin{Bmatrix} \ddot{x}_1 \\ \ddot{x}_2 \end{Bmatrix}_{+\alpha2} \begin{bmatrix} 4 & -2 \\ -2 & 2 \end{bmatrix} \begin{Bmatrix} x_1 \\ x_2 \end{Bmatrix}_{=\alpha2} \begin{Bmatrix} Y_0 \ sin \ \omega t \\ 0 \end{Bmatrix}$$

where $\quad \alpha^2 = \dfrac{12 \ EI}{mH^3} = \dfrac{12}{m} = 6$

Solving for steady-state response we get:

$$X_1 = \frac{(\alpha^2 - \omega^2) \alpha^2}{(\omega^2 - \omega_1^2)(\omega^2 - \omega_2^2)} Y_0$$

$$X_2 = \frac{\alpha^4}{(\omega^2 - \omega_1^2)(\omega^2 - \omega_2^2)} Y_0$$

These values are to be plotted against various values of ω.

Solution:

The MATLAB script for this problem is given as follows:

```
m=20;% mass
k=200; % k=12EI/H3  stiffness
w0=k/m;
M=[m 0;0 m]; %mass matrix
K=[2*k -k;-k k]; % stiffness matrix
%eigenvalues
[u,W]=eig(K,M);
u(:,1)=u(:,1)/max(u(:,1));
u(:,2)=u(:,2)/max(u(:,2));
```

```
[wn(1),I1]=min(max(W));
[wn(2),I2]=max(max(W));
wn(1)=sqrt(wn(1)); % Nat. frequency 1
wn(2)=sqrt(wn(2)); % Nat. frequency 2
U(:,1)=u(:,I1);
U(:,2)=u(:,I2);
w=[0:0.002:6];
T2=(w0^2)./((w.^2-wn(1)^2).*(w.^2-wn(2)^2));
plot(w,T2)
title('Frequency Response')
ylabel('{\it{X}}_2(\omega)/{\it{Y}}_0')
xlabel('\omega')
axis([0 8 -5 5])
grid
```

The MATLAB output is shown in Fig. E3.17(a).

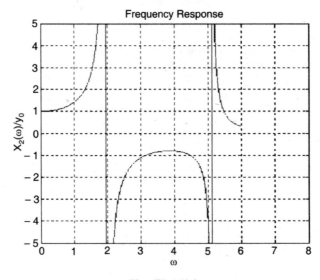

Fig. E3.17(a)

Example 3.18. *A 3-degree of freedom system shown in Fig. E3.18. Obtain the natural frequencies and mode shapes using MATLAB script. Assume $k = m = 1$.*

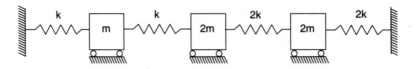

Fig. E3.18

The equations of motion can be written as:

$$M\ddot{x}(t) + Kx(t) = 0$$

with $x(t) = [x_1(t)\ x_2(t)\ x_3(t)]T$ as the displacement vector

$$M = \begin{bmatrix} m & 0 & 0 \\ 0 & 2m & 0 \\ 0 & 0 & 2m \end{bmatrix} \text{ and } K = \begin{bmatrix} 2k & -k & 0 \\ -k & 3k & -2k \\ 0 & -2k & 4k \end{bmatrix}$$

as the mass and stiffness matrices.

Solution:

The MATLAB script for finding the natural frequencies and mode shapes is given as follows:

```
k=1; % stiffness
m=1; % mass
M=m*[1 0 0;0 2 0;0 0 2]; % mass matrix
K=k*[2 -1 0;-1 3 -2;0 -2 4]; % stiffness matrix
N=3;
R=chol(M); % Cholesky decomposition technique
L=R';
A=inv(L)*K*inv(L');
[x,W]=eig(A);
v=inv(L')*x;
for i=1:N,
   w1(i)=sqrt(W(i,i));
end
[w,I]=sort(w1);
disp('The first three natural frequencies are')
disp(w(1))
disp(w(2))
disp(w(3))
n=[1:N];
disp('The corresponding mass-orthonormalized mode shapes are')
for j=1:N,
   U(:,j)=v(:,I(j));
   U(:,j)=U(:,j)/(U(:,j)'*M*U(:,j));
   disp('mode-')
   disp(j)
   disp(U(:,j))
end
```

The outputs are as follows:

The first three natural frequencies are

```
    0.7071
    1.4142
    1.7321
```

The corresponding mass-orthonormalized mode shapes are

```
mode-
    1
  -0.3651
  -0.5477
```

```
    -0.3651
mode-
     2
   0.8165
  -0.0000
  -0.4082
mode-
     3
  -0.4472
   0.4472
  -0.4472
```

Example 3.19. *In Fig. E3.18, if mass m is subjected to unit step function u(t), determine the response using modal analysis. Write a MATLAB script to plot the displacement response of all the masses.*

Solution:

The MATLAB program is given as follows:

```
M=[1 0 0;0 2 0;0 0 2]; % mass matrix
C=[0 0 0;0 0 0;0 0 0]; % damping matrix
K=[2 -1 0;-1 3 -2;0 -2 4]; % stiffness
A=[zeros(size(M)) eye(size(M));-inv(M)*K -inv(M)*C];
B=[zeros(size(M)); inv(M)];
TO=10; %RISE TIME OF FORCE
N=200;
T=0.1; % SAMPLING PERIOD
NO=TO/T;
phi=eye(size(A))+T*A+T^2*A^2/2+T^3*A^3/6;
gamma=inv(A)*(phi-eye(size(A)))*B;
x(:,1)=zeros(2*length(M),1);
for k=1:N,
   f(k)=1;
F(:,k)=[1;0;0]*f(k); % Force is only applied to mass m
x(:,k+1)=phi*x(:,k)+gamma*F(:,k);
end
k=[0:N];
plot(k,x(1,:),'o',k,x(2,:),'s',k,x(3,:),'.')
title('system response for unit step at first mass E3.19')
ylabel('x_1(k),x_2(k),x_3(k)')
xlabel('k')
legend('x_1(k)','x_2(k)','x_3(k)')
grid
```

The output obtained is shown in Fig.E3.19(a).

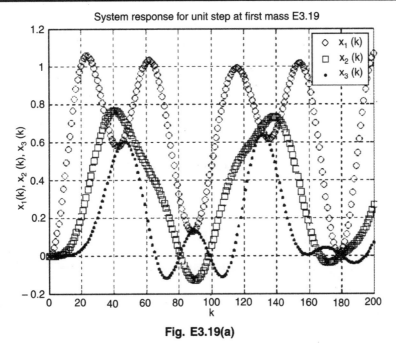

Fig. E3.19(a)

Example 3.20. *A two-degree of freedom torsional system shown in Fig. E3.20 and is subjected to a torque of unit pulse nature [u(t) – u(t – 4)] at the disc B.*

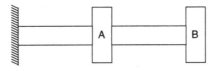

Fig. E3.20

The mass, stiffness and damping matrices are

$$M = \begin{bmatrix} 3 & 0 \\ 0 & 5 \end{bmatrix}, K = \begin{bmatrix} 5 & -4 \\ -4 & 4 \end{bmatrix} \text{ and } C = \begin{bmatrix} 1.6 & -0.8 \\ -0.8 & 0.8 \end{bmatrix}$$

Plot the response of disc A using MATLAB.

Solution:

The MATLAB script is given as follows:

```
M=[1 0;0 2]; % mass matrix
C=[1.6 -0.8; -0.8 0.8]; % damping matrix
K=[5 -4;-4 4]; % stiffness matrix
A=[zeros(size(M)) eye(size(M));-inv(M)*K  -inv(M)*C];
B=[zeros(size(M)); inv(M)];
TO=4; %RISE TIME OF FORCE
N=600;
T=0.1; % SAMPLING PERIOD
NO=TO/T;
phi=eye(size(A))+T*A+T^2*A^2/2+T^3*A^3/6;
gamma=inv(A)*(phi-eye(size(A)))*B;
```

```
x(:,1)=zeros(2*length(M),1);
for k=1:N,
   if k<=N0+1; f(k)=1;
   else;f(k)=0;end
   F(:,k)=[0;1]*f(k); % Force is only applied to mass m
   x(:,k+1)=phi*x(:,k)+gamma*F(:,k);
end
k=[0:N];
plot(k,x(1,:),'.')
title('system response for unit step at first disc E3.20')
ylabel('x_1(k)')
xlabel('k')
grid
```

The output of this program is given in Fig. E3.20(a).

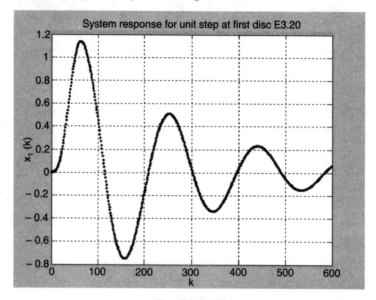

Fig. E3.20(a)

Example 3.21. *For the single degree of freedom vibrating system shown in Fig. E3.21, determine the motion of the mass subjected to the initial conditions $x(0) = 0.15$ m and $\dot{x} = 0.04$ m/s. Given $m = 1$ kg, $c = 5$ N-s/m, and $k = 5$ N/m.*

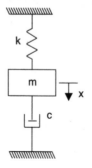

Fig. E3.21

Solution:

The system equation is

$$m\,\ddot{x} + c\,\dot{x} + kx = 0$$

with the initial conditions $x(0) = 0.15$ m and $\dot{x} = 0.04$ m/s. The Laplace transform of the system equation gives

$$m[s^2X(s) - sx(0) - \dot{x}(0)] + c[sX(s) - x(0)] + kX(s) = 0$$

or

$$(ms^2 + cs + k)X(s) = mx(0)s + m\,\dot{x}(0) + cx(0)$$

Solving this last equation for $X(s)$ and substituting the given numerical values, we obtain

$$X(s) = \frac{mx(0)\,s + m\dot{x}(0) + cx(0)}{ms^2 + cs + k} = \frac{0.15s + 0.79}{s^2 + 5s + 5}$$

This equation can be written as

$$X(s) = \frac{0.15s^2 + 0.79s}{s^2 + 5s + 5}\,\frac{1}{s}$$

Hence the motion of the mass m may be obtained as the unit-step response of the following system:

$$G(s) = \frac{0.15s + 0.79}{s^2 + 5s + 5}$$

MATLAB program will give a plot of the motion of the mass. The plot is shown in Fig. E3.21(a).

```
num = [0.15    0.79    0];
den = [1    5    5];
step(num,den)
grid
title('Response of spring mass-damper system to initial condition')
```

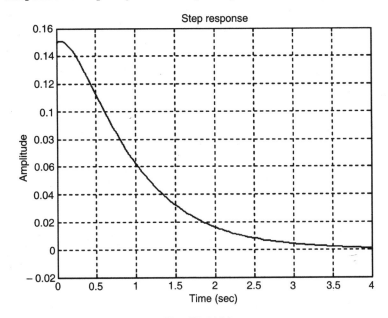

Fig. E3.21(a)

Example 3.22. *For the vibrating single degree of freedom shown in Fig.E3.22, determine the response of the system when 12 N of free (step input) is applied to the mass m and plot the response using MATLAB. Given that the system is at rest initially and the displacement x is measured from the equilibrium position. Assume that m = 2 kg, c = 10 N-s/m, and k = 80 N/m.*

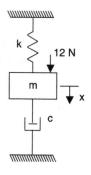

Fig. E3.22

Solution:

The equation of motion for the system is

$$m\ddot{x} + c\dot{x} + kx = P$$

By substituting the numerical values into this last equation, we get

$$2\ddot{x} + 10\dot{x} + 80x = 12$$

By taking the Laplace transform of this last equation and substituting the initial conditions [$x(0) = 0$ and $\dot{x}(0) = 0$], the result is

$$(s^2 + 5s + 40)X(s) = \frac{6}{s}$$

Solving for $X(s)$, we obtain

$$X(s) = \frac{6}{s\,(s^2 + 5s + 40)}$$

The response exhibits damped vibrations.

MATLAB program is used to a plot of the response curve, which is shown in Fig. E3.22(a).

```
num = [0    0    6];
den = [1    5    40];
step(num,den)
grid
```

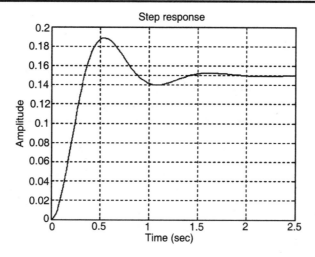

Fig. E3.22(a)

Example 3.23. *For the mechanical system shown in Fig. E3.23, obtain the response $x_0(t)$ when $x_i(t)$ is a unit step displacement input. Assume that $k_1 = 2 \ N/m$, $k_2 = 4 \ N/m$, $c_1 = 1 \ N\text{-}s/m$, and $c_2 = 2 \ N\text{-}s/m$.*

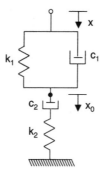

Fig. E3.23

Solution:

The transfer function $X_0(s)/X_i(s)$ is given by

$$\frac{X_0 \, (s)}{X_i \, (s)} = \frac{\left(\dfrac{c_1}{k_1} s + 1\right)\left(\dfrac{c_2}{k_2} s + 1\right)}{\left(\dfrac{c_1}{k_1} s + 1\right)\left(\dfrac{c_2}{k_2} s + 1\right) + \dfrac{c_2}{k_1} s}$$

Substitution of the given numerical values yields

$$\frac{X_0 \, (s)}{X_i \, (s)} = \frac{(0.5s + 1)(0.5s + 1)}{(0.5s + 1)(0.5s + 1) + s} = \frac{0.25s^2 + s + 1}{0.25s^2 + 2s + 1} = \frac{s^2 + 4s + 4}{s^2 + 8s + 4}$$

The MATLAB program is used to obtain the unit-step response is given below:

```
num = [1    4    4];
den = [1    8    4];
step(num,den)
grid
```

The output is shown as in Fig. E3.23 (*a*)

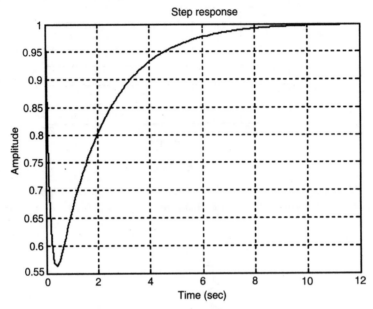

Fig. E3.23(a)

Example 3.24. *The impulse response of a second-order system is given as*

$$\frac{C(s)}{R(s)} = \frac{\omega_n^2}{s^2 + 2\xi\omega_n s + \omega_n^2}$$

For a unit-impulse input R(s) = 1, and ω_n *= 1 rad / sec, C(s) is given by*

$$C(s) = \frac{1}{s^2 + 2\xi s + 1}$$

Plot the ten unit-impulse response curves in one diagram using MATLAB for ξ *= 0.1, 0.2, 0.3, 0.4, 0.5, 0.6, 0.7, 0.8, 0.9, and 1.0.*

Solution:

The MATLAB program:

```
num = [0    0    1];
den1 = [1    0.2    1];
t = 0:0.1:10;
impulse(num,den1,t);
text(2.2, 0.88, 'Zeta = 0.1')
hold
current plot held
den2 = [1    0.4    1]; den3 = [1    0.6    1]; den4 = [1    0.8    1];
den5 = [1    1    1]; den6 = [1    1.2    1]; den7=[1    1.4    1];
den8 = [1    1.6    1]; den9 = [1    1.8    1]; den10 = [1    2.0    1];
```

```
impulse(num,den2,t)
text(1.3,0.7,'0.3')
impulse(num,den3,t)
text(1.15,0.58,'0.5')
impulse(num,den4,t)
text(1.1,0.46,'0.7')
impulse(num,den5,t)
text(0.8,0.38,'1.0')
impulse(num,den6,t)
text(0.7,0.28,'1.0')
impulse(num,den7,t)
text(0.6,0.24,'1.0')
impulse(num,den8,t)
text(0.5,0.21,'1.0')
impulse(num,den9,t)
text(0.4,0.18,'1.0')
impulse(num,den10,t)
text(0.3,0.15,'1.0')
grid
title('Impulse-response curve for G(s) = 1/[s^2+2(zeta)s+1]')
hold
current plot released
```

The output is shown in Fig. E3.24 (*a*)

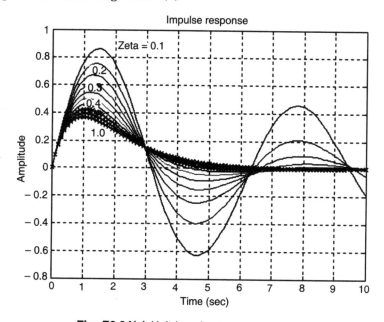

Fig. E3.24(a) Unit-impulse response curves.

Example 3.25. *For the mechanical system shown in Fig. E3.25, assume that m = 1 kg, m_1 = 2kg, k_1 = 15 N/m, and k_2 = 60 N/m. Determine the vibration when the initial conditions are given as: x(0) = 0.23 m, $\dot{x}$ = 0 m/s, y(0) = 1 m, $\dot{y}$ = 0 m/s. Write a MATLAB program to plot curves x(t) versus t and y(t) versus t for the initial conditions.*

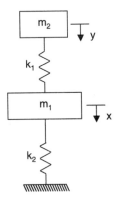

Fig. E3.25

Solution:

The equations for the system are

$$(2s^2 + 75)X(s) = 2sx(0) + 15Y(s) \tag{1}$$
$$(s^2 + 15)Y(s) = sy(0) + 15X(s) \tag{2}$$

Solving we obtain

$$X(s) = \frac{2(s^2 + 15)\,sx\,(0) + 15\,sy\,(0)}{2s^4 + 105s^2 + 900} \tag{3}$$

For the initial conditions

$$x(0) = 0.23 \text{ m}, \ \dot{x} = 0 \text{ m/s}, \ y(0) = 1 \text{ m}, \ \dot{y}(0) = 0 \text{ m/s}$$

Eq. (3) becomes as follows:

$$X(s) = \frac{0.46s^3 + 21.9s}{2s^4 + 105s^2 + 900}$$

$$= \frac{0.46s^4 + 21.9s^2}{2s^4 + 105s^2 + 900} \frac{1}{s} \tag{4}$$

By substituting Eq.(4) into Eq.(2) and solving for Y(s), we obtain

$$Y(s) = \frac{1}{s^2 + 15}\,sy(0) + 15X(s)$$

Substituting y(0) = 1 into the last equation and simplifying, we get

$$Y(s) = \frac{1}{s^2 + 15}\,\frac{24s^5 + 105s^3 + 900s + 6.9s^3}{2s^4 + 105s^2 + 900}\,\frac{328.5s}{}$$

To obtain plot of x(t) versus t, we may enter the following MATLAB program into the computer. The resulting plots are shown in Fig.E3.25(a). Likewise y(t) versus t can be also plotted.

```
num1 = [0.46   0   21.9   0   0];
den = [2    0   105   0 900];      % see equation (4)
t=0:0.01:20;
x=step(num1,den,t)

plot(t,x)
title('Responses mass m1- x(t) due to initial conditions')
xlabel('t sec')
ylabel('x(t)')
grid
```

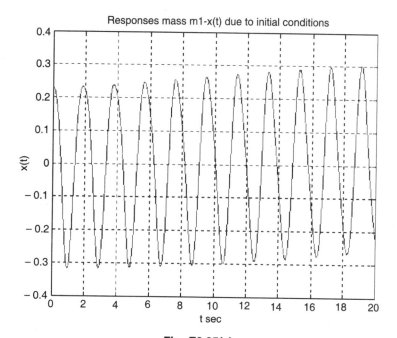

Fig. E3.25(a)

Example 3.26. *For the mechanical system shown in Fig.E3.25, assume that m = 1 kg,*
$m_1 = 2kg$, $k_1 = 15$ N/m, *and* $k_2 = 60$ N/m. *Determine the vibration when the initial conditions are*
given as: $x(0) = 1.75$ m, $\dot{x}(0) = 0$ m/s, $y(0) = -1$ m, $\dot{y}(0) = 0$ m/s. *Write a MATLAB program to plot*
curves x(t) versus t and y(t) versus t for the initial conditions.

Solution:

$$X(s) = \frac{2(s^2 + 15)sx(0) + 15sy(0)}{2s^4 + 105s^2 + 900}$$

$$Y(s) = \frac{1}{s^2 + 15}sy(0) + 15 X(s)$$

For the initial conditions

$$x(0) = 1.75 \text{ m}, \dot{x}(0) = 0 \text{ m/s}, y(0) = -1 \text{ m}, \dot{y}(0) = 0 \text{ m/s}$$

we obtain the following expressions for $X(s)$ and $Y(s)$:

$$X(s) = \frac{3.5s^3 + 37.5s}{2s^4 + 105s^2 + 900} = \frac{3.5s^4 + 37.5s^2}{2s^4 + 105s^2 + 900} \frac{1}{s}$$

$$Y(s) = \frac{1}{s^2 + 15} - s + 15X(s)$$

A MATLAB program for obtaining plots of $x(t)$ versus t given below. The resulting plot is shown in Fig. E3.26.

```
num1 = [3.5   0 37.5 0   0];
den = [2    0   105 0 900];
step(num1,den)
```

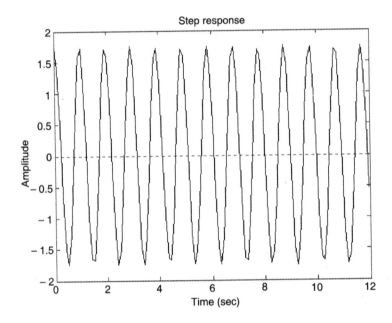

Fig. E3.26 Plot of motion of mass m_1.

Example 3.27. *For the mechanical system shown in Fig. E3.25, assume that m = 1 kg, m_1 = 2kg, k_1 = 15 N/m, and k_2 = 60 N/m. Determine the vibration when the initial conditions are given as: x(0) = 0.5 m, $\dot{x}(0)$ = 0 m/s, y(0) = – 0.5 m, $\dot{y}(0)$ = 0 m/s. Write a MATLAB program to plot curves x(t) versus t and y(t) versus t for the initial conditions.*

Solution:

$$X(s) = \frac{2(s^2 + 15)\, sx(0) + 15sy(0)}{2s^4 + 105s^2 + 900}$$

$$Y(s) = \frac{1}{s^2 + 15}\, sy(0) + 15X(s)$$

For the initial conditions

$$x(0) = 0.5 \text{ m}, \ \dot{x}(0) = 0 \text{ m/s}, \ y(0) = -0.5 \text{ m}, \dot{y}(0) = 0 \text{ m/s}$$

we obtain the following expressions for $X(s)$ and $Y(s)$:

$$X(s) = \frac{s^4 + 7.5s^2}{2s^4 + 105s^2 + 900} \frac{1}{s}$$

$$Y(s) = \frac{-0.5s}{s^2 + 15} \frac{15X(s)}{s^2 + 15}$$

A MATLAB program for obtaining plots of $x(t)$ versus t given below. The resulting plots are shown in Fig. E3.27. Likewise $y(t)$ can also be plotted

```
num1 = [1   0   7.5   0   0];
den = [2    0    105   0   900];
t = 0:0.02:5;
x = step(num1,den,t)
plot(t, x, 'o')
title('Responses x(t) due to initial conditions ')
xlabel('t sec')
ylabel('x(t)')
grid
```

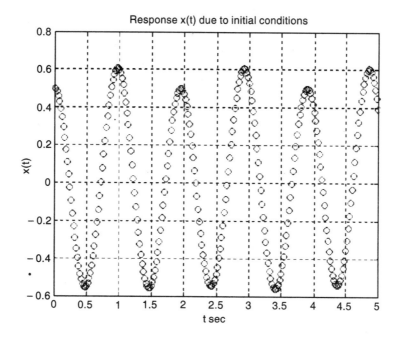

Fig. E3.27

3.3 SUMMARY

With this foundation of basic application of MATLAB, this Chapter provides opportunities to explore advanced topics in vibration analysis engineering. Extensive worked examples are included with a significant number of exercise problems to guide the student to understand and as an aid for learning about the vibration analysis of mechanical systems using MATLAB.

PROBLEMS

P3.1 A safety bumper placed at the end of a race track to stop out-of-control cars as shown in
Fig. P3.1. The bumper is designed such that the force that the bumper applies to the car
is a function of the velocity v and the displacement x of the front end of the bumper
given by the equation:

$$F = Kv^3 (x + 1)^3$$

where $K = 35$ kg-s/m^5 (a constant).

A car with a mass of 2000 kg hits the bumper at a speed of 100 km/h. Determine and plot
the velocity of the car as a function of its position for $0 \le x \le 5$ m.

Fig. P3.1

P3.2 The 10 kg body is moved 0.25 m to the right of the equilibrium position and released
from rest at $t = 0$ as shown in Fig. P3.2. Plot the displacement as a function of time for
four cases: $c = 10, 40, 50$ and 60 N.s/m. The stiffness of the spring is 40 N/m.

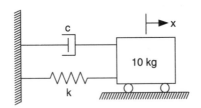

Fig. P3.2

P3.3 An airplane uses a parachute and other means of braking as it slows down on the run-
way after landing. The acceleration of the airplane is given by

$$a = - 0.005 \, v^2 - 4 \text{ m/s}^2$$

Consider an airplane with a velocity of 500 km/h that opens its parachute and starts
decelerating at $t = 0$ s.

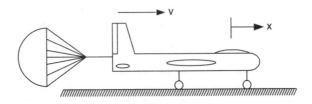

Fig. P3.3

P3.4 The piston of 150 *lb* is supported by a spring of modulus k = 250 lb/in. A dashpot of damping coefficient c = 100 lb.sec/ft acts in parallel with the spring. A fluctuating pressure p = 0.75 sin 30t (psi) acts on the piston, whose top surface area is 100 in². Plot the response of the system for initial conditions x_0 = 0.06 ft and $\dot{x}_0$ = 6, 0, and − 6 ft./sec.

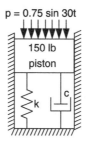

Fig. P3.4

P3.5 The 15 kg oscillator contains an unbalanced motor whose speed is N rpm as shown in Fig. P3.5. The stiffness of the spring k = 1100 N/m. The oscillator is also restrained by a viscous damper whose piston is resisted by a force of 50 N when moving at a speed of 0.6 m/s. Determine,

(*a*) the viscous damping factor

(*b*) plot the magnification factor for motor speeds from 0 to 350 rpm

(*c*) the maximum value of the magnification factor and the corresponding motor speed.

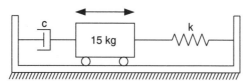

Fig. P3.5

P3.6 Write a MATLAB script file that computes the response of a single degree of freedom under damped system shown in Fig. P3.6 to initial excitations. Use the program to determine and plot the response for the following data:

Initial conditions:

$$x(0) = 0, \quad \dot{x}(0) = v_0 = 30 \text{ cm./sec},$$
$$\omega_n = 6 \text{ rad/s, and } \xi = 0.05, 0.1, 0.2, 0.30.$$

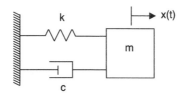

Fig. P3.6 Damped single degree of freedom system.

The response of the under damped single degree of freedom system is given by

$$x(t) = A \, e^{-\xi\omega_n t} \cos(\omega_d t - \phi)$$

where A and ϕ represent the amplitude and phase angle of the response respectively. There are

$$A = \sqrt{x_0^2 + \left(\frac{\zeta\omega_n x_0 + v_0}{\omega_d} \right)^2}$$

$$\omega_d = \sqrt{1 - \zeta^2} \; \omega_n$$

and

$$\phi = \tan^{-1} \left(\frac{\zeta\omega_n x_0 + v_0}{\omega_d x_0} \right)$$

P3.7 Write a MATLAB script for plotting the frequency response magnitude and phase angle using complex notation for a single degree of freedom system given by

$$G(i\omega) = \frac{1 - \left(\dfrac{\omega}{\omega_n} \right)^2 - i \, 2\zeta \left(\dfrac{\omega}{\omega_n} \right)}{\left[1 - \left(\dfrac{\omega}{\omega_n} \right)^2 \right] + \left(2\zeta \dfrac{\omega}{\omega_n} \right)^2}$$

and

$$\phi(\omega) = \tan^{-1} \left[\frac{-\operatorname{Im} G(i\omega)}{\operatorname{Re} G(i\omega)} \right] = \tan^{-1} \left[\frac{2\zeta \dfrac{\omega}{\omega_n}}{1 - \left(\dfrac{\omega}{\omega_n} \right)^2} \right]$$

P3.8 Consider the force-free, viscously damped single degree of freedom system shown in Fig. P3.8.

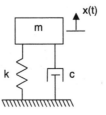

Fig. P3.8

Plot the response of the system using MATLAB over the interval $0 \le t \le 10s$ to the initial conditions $x(0) = 3$ cm, $\dot{x} = 0$ for the values of the damping factor $\zeta = 0.05, 0.1, 0.5$. The frequency of the undamped oscillation have the values $\omega_n = 15$ rad/s.

The expression for the response of an damped single degree of freedom system in Fig.P3.3 to initial displacement and velocity is given by

$$x(t) = C e^{-\zeta\omega_n t} \cos(\omega_d t - \phi)$$

where C and ϕ represent the amplitude and phase angle of the response, respectively having the values

$$C = \sqrt{x_0^2 + \left(\frac{\zeta\omega_n x_0 + v_0}{\omega_d}\right)^2}$$

$$\phi = \tan^{-1}\left(\frac{\zeta\omega_n x_0 + v_0}{\omega_d x_0}\right)$$

and $\qquad \omega_d = \sqrt{1-\zeta^2}\ \omega_n$

P3.9 Write a MATLAB script to obtain the motion of the mass subjected to the initial condition. There is no external forcing function acting on the system. The single degree of freedom system is shown in Fig. P3.9 and the parameters are given as $m = 3$ kg, $k = 6$ N/m, and $C = 5$ N-s/m. The displacement of the mass is measured from the equilibrium position and at $t = 0$, $x(0) = 0.04$ m and $\dot{x}(0) = 0.10$ m/s.

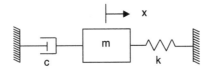

Fig. P3.9 Single degree of freedom system.

P3.10 Determine and plot the response of the single degree of freedom system shown in Fig. P3.10 using MATLAB when 25 N of force (step input) is applied to the mass m. The system is at rest initially and the displacement of the mass m is measured from equilibrium position. The parameters of the system are given as $m = 3$ kg, $c = 25$ N-s/m, and $k = 200$ N/m. The initial conditions are $x(0) = \dot{x}(0) = 0$.

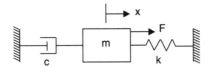

Fig. P3.10 Single degree of freedom system.

P3.11 Write a MATLAB script for determining the response of a single degree of freedom system with viscous damping to an exponential excitation $F(t) = e^{-\alpha t}$.

P3.12 A single degree of freedom spring-mass-damper model has following properties: $m = 15$ kg, $c = 25$ Ns/m and $k = 3500$ N/m. If it is subjected to a triangular pulse of amplitude 1000 N for 0.1 seconds, compute the time-domain response and plot the same in MATLAB.

The excitation function is shown in Fig. P3.12.

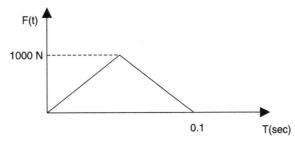

Fig. P3.12

P3.13 Determine and plot the response of the system shown in Fig. P3.13 using MATLAB. The response is $x_0(t)$ when the input $x_i(t)$ is a unit step displacement input. The parameters of the system are $k_1 = 15$ N/m, $k_2 = 25$ N/m, $c_1 = 7$ N-s/m, $c_2 = 15$ N-s/m.

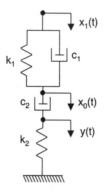

Fig. P3.13

P3.14 A two-degree of freedom torsional system shown in Fig. P3.14 is subjected to initial excitation $\theta_1(0) = 0$, $\theta_2(0) = 2$, $\dot{\theta}_1(0) = 2\sqrt{\dfrac{GJ}{IL}}$ and $\dot{\theta}_2(0) = 0$. Write MATLAB program and plot the response of the system. Assume $I = 1$ and $GJ = l = 1$.

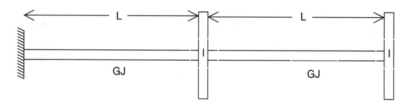

Fig. P3.14

P3.15 The mass m_2 in a 2-degree of freedom system shown in Fig. P3.15 is subjected to a force in the form of saw-tooth pulse of amplitude 1.5 N for duration of 1.5 second. Obtain the response in terms of two coordinates $x_1(t)$ and $x_2(t)$. Assume $k_1 = k_2 = 15$N/m and $m_1 = m_2 = 2$kg.

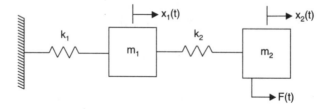

Fig. P3.15

The mass and stiffness matrices of the system for given system are given as

$$M = \begin{bmatrix} m_1 & 0 \\ 0 & m_2 \end{bmatrix} \text{ and } K = \begin{bmatrix} k_1 + k_2 & -k_2 \\ -k_2 & k_2 \end{bmatrix}$$

Force vector is:

$$F = \left\{ \begin{matrix} F(t) \\ 0 \end{matrix} \right\}$$

The saw-tooth pulse takes the form as shown in Fig. P3.15(a).

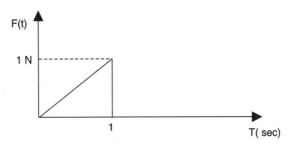

Fig. P3.15(a)

P3.16 A two-story building (Fig. P3.16) is undergoing a horizontal motion $y(t) = Y_0 \sin \omega t$.

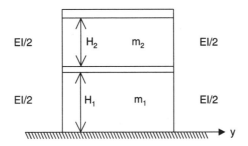

Fig. P3.16

Derive expression for the displacement of the first floor having mass m_1. Assume $m_1 = m_2 = 4$, $EI = 2$ and $H = 1$ m.

The equations of motion for building can be written as:

$$\begin{bmatrix} 4 & 0 \\ 0 & 4 \end{bmatrix} \left\{ \begin{matrix} \ddot{x}_1 \\ \ddot{x}_2 \end{matrix} \right\} + \alpha^2 \begin{bmatrix} 2 & -1 \\ -1 & 1 \end{bmatrix} \left\{ \begin{matrix} x_1 \\ x_2 \end{matrix} \right\} = \alpha^2 \left\{ \begin{matrix} Y_0 \sin \omega t \\ 0 \end{matrix} \right\}$$

where $\quad \alpha^2 = \dfrac{12EI}{mH^3} = \dfrac{12 \times 2}{m} = 6$

Solving for steady-state response we get:

$$X_1 = \frac{(\alpha^2 - \omega^2)\alpha^2}{(\omega^2 - \omega_1^2)(\omega^2 - \omega_2^2)} Y_0$$

$$X_2 = \frac{\alpha^4}{(\omega^2 - \omega_1^2)(\omega^2 - \omega_2^2)} Y_0$$

These values are to be plotted against various values of ω.

P3.17 Derive the response of the system shown in Fig. P3.17 in discrete time and plot the response. Given $F(t) = e^{-\alpha t}$.

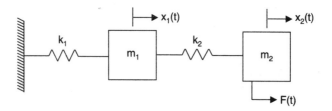

Fig. P3.17

P3.18 Consider the system with $M = \begin{bmatrix} 3 & 0 \\ 0 & 2 \end{bmatrix}$, $K = \begin{bmatrix} 6 & -4 \\ -4 & 5 \end{bmatrix}$ with arbitrary viscous damping.

Find the eigenvalues and normalized eigenvectors.

P3.19 For the vibrating system shown in Fig. E3.19, a mass of 5 kg is placed on mass m at $t = 0$ and the system is at rest initially (at $t = 0$). Given that $m = 20$ kg, $k = 600$ N/m, and $c = 60$ Ns/m. Plot the response curve $x(t)$ versus t using MATLAB.

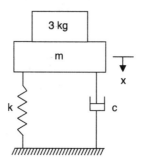

Fig. P3.19

P3.20 For the mechanical vibrating system shown in Fig.P3.20, using MATLAB assume that $m = 3$ kg, $k_1 = 15$ N/m, $k = 25$ N/m, and $c = 10$ N-s/m. Plot the response curve $x(t)$ versus t when the mass m is pulled slightly downward and the initial conditions are $x(0) = 0.05$ m and $\dot{x} = 0.8$ m/s.

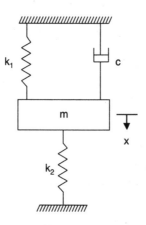

Fig. P3.20

P3.21 For the mechanical vibrating system shown in Fig. P3.21, $k_1 = 10$ N/m, $k_2 = 30$ N/m, $c_1 = 3$ N-s/m, and $c_2 = 25$ N-s/m.

(a) Determine the displacement $x_2(t)$ when F is a step force input of 4 N.

(b) Plot the response curve $x_2(t)$ versus t using MATLAB.

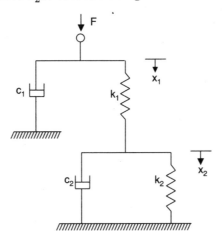

Fig. P3.21

P3.22 For the electrical system shown in Fig. E3.22, assume that $R_1 = 2$ Ω, $R_2 = 1$ $M\Omega$, $C_1 = 0.75$ μF, and $C_2 = 0.25$ μF and the capacitors are not charged initially and $e_0(0) = 0$ and $\dot{e}_0(0) = 0$.

(a) Find the response $e_0(t)$ where $e_t(t) = 5$ V (stop input) is applied to the system.

(b) Plot the response curve $e_0(t)$ versus t using MATLAB.

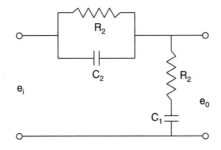

Fig. P3.22

P3.23 For the mechanical system shown in Fig. P3.23, assume $m = 3$ kg, $M = 25$ kg, $k_1 = 25$ N/m, and $k_2 = 300$ N/m. Determine

(a) the natural frequencies and modes of vibration

(b) the vibration when the initial conditions are: $x(0) = 0.05$ m, $\dot{x}(0) = 0$ m/s, $y(0) = 0$ m,
and $\dot{y}(0) = 0$ m/s.

Use MATLAB program to plot curves $x(t)$ versus t and $y(t)$ versus t.

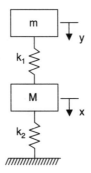

Fig. P3.23

Bibliography

There are several outstanding text and reference books on vibration analysis, numerical methods, and MATLAB that merit consultation for those readers who wish to pursue these topics further. Also, there are several publications devoted to presenting research results and in-depth case studies in vibration analysis. The following list is but a representative sample of the many excellent references that includes journals and periodicals on vibration analysis, numerical methods, and MATLAB.

Adams, M.L., *Rotating Machinery Vibration*, Marcel Dekker, New York, NY, 2002.

Anderson, J.F., and Anderson, M.B., *Solution of Problems in Vibrations,* Longman Scientific and Technical, Essex, UK, 1987.

Anderson, R.A., *Fundamentals of Vibrations, Macmillan*, New York, NY, 1967.

Balachandran, B., and Magrab, E.B., *Vibrations*, Brooks/Cole, Pacific Grove, CA, 2004.

Barker, J.R., *Mechanical and Electrical Vibrations*, Wiley, New York, NY, 1964.

Beards, C.F., *Structural Vibration Analysis,* Ellis Harwood, U.K, 1983.

Beards, G.F., *Vibrations and Control System,* Ellis Harwood, UK, 1988.

Benaroya, H., *Mechanical Vibrations*, Prentice Hall, Upper Saddle River, NJ, 1998.

Bendat, J.S., and Piersol, A.G., *Engineering Applications of Correlation and Spectral Analysis*, Wiley, New York, 1980.

Bendat, J.S., and Piersol, A.G., *Measurement and Analysis of Random Vibration Data*, Wiley, New York, NY, 1965.

Bendat, J.S., and Piersol, A.G., *Random Data*, Wiley, New York, NY, 1986.

Bendat, J.S., and Piersol, A.G., *Random Data: Analysis and Measurement Procedures*, Wiley, New York, NY, 1971.

Beranek, L.L., and Ver, I.L., *Noise and Vibration Control Engineering: Principles and Applications*, Wiley, New York, NY, 1992.

Beranek, L.L., *Noise and Vibration Control*, McGraw Hill, New York, NY, 1971.

Berg, G.V., *Elements of Structural Dynamics*, Prentice Hall, Englewood cliffs, NJ, 1989.

Bernhard, R.K., *Mechanical Vibrations*, Pitman Publishing, 1943.

Bhat, R.B., and Dukkipati, R.V., *Advanced Dynamics,* Narosa Publishing House, New Delhi, India, 2001.

Bickley, W.G., and Talbot, A., *Vibrating Systems*, Oxford University Press, Oxford, 1961.

Bishop, R. E.D., *Vibration,* Cambridge University Press, Cambridge, England, 1979.

Bishop, R.E.D., and Gladwell, G.M.L., *The Matrix Analysis of Vibration*, Cambridge University Press, Cambridge, England, 1965.

Bishop, R.E.D., and Johnson, D.C., *Vibration Analysis Tables*, Cambridge University Press, Cambridge, England, 1956.

Bishop, R.F.D., and Johnson, D.C., *The Mechanics of Vibration*, Cambridge University Press, New York, NY, 1960.

Blevins, R.D., *Formulas for Natural Frequencies and Mode Shapes*, R.E. Krieger, Melbourne, FL, 1987.

Broch, J.F., *Mechanical Vibrations and Shock Measurements*, Larson & Sons, Copenhagen, Denmark, 1980.

Brommundt, E., *Vibration of Continuous Systems*, CISM, Udine, Italy, 1969.

Burton, R., *Vibration and Impact,* Dover Publications, New York, NY, 1958.

Bykhovsky, I., *Fundamentals of Vibration Engineering*, MIR Publications, 1972.

Centa, G., *Vibration of Structures and Machines,* Springer Verlag, New York NY, 1993.

Chen, Y., *Vibrations*: *Theoretical Methods*, Addison-Wesley, Reading, MA, 1966.

Church, A.H., *Mechanical Vibrations,* 2nd ed., Wiley, New York, NY, 1963.

Cole, E.B., *The Theory of Vibrations for Engineers*, Crosby Lockwood, 1950.

Crafton, P.A., *Shock and Vibration in Linear Systems,* Harper & Row, New York, NY, 1961.

Crandall, S.H., and Mark, W.D., *Random Vibration in Mechanical Systems*, Academic Press, New York, NY, 1963.

Crandall, S.H., *Random Vibration*, MIT Press, Cambridge, MA, 1963.

De Silva, C.W., *Vibration: Fundamentals and Practice*, CRC Press, Boca Raton, FL, 2000.

Del Pedro, M., and Pahud, P., *Vibration Mechanics,* Kluwer Academic Publishers, Dordrecht, Netherlands, 1989.

Den Hartog. J.P., *Mechanical Vibrations*, 4th ed., McGraw Hill, New York, NY, 1956.

Dimarogonas, A.D., and Haddad, S.D., *Vibration for Engineers,* Prentice Hall, Englewood cliffs, NJ, 1992.

Dimarogonas, A.D., *Vibration for Engineers,* 2nd ed., Prentice Hall, Englewood cliffs, NJ, 1996.

Dukkipati, R.V., *Advanced Engineering Analysis***,** Narosa Publishing House, New Delhi, India, 2006.

Dukkipati, R.V., *Advanced Mechanical Vibrations***,** Narosa Publishing House, New Delhi, India, 2006.

Dukkipati, R.V., and Amyot, J.R., *Computer Aided Simulation in Railway Vehicle Dynamics,* Marcel-Dekker, New York, NY, 1988.

Dukkipati, R.V., and Srinivas, J., *A Text Book of Mechanical Vibrations,* Prentice Hall of India, New Delhi, India, 2005.

Dukkipati, R.V., and Srinivas, J., *Vibrations: Problem Solving Companion,* Narosa Publishing House, New Delhi, India, 2006.

Dukkipati, R.V., *Vehicle Dynamics,* Narosa Publishing House, New Delhi, India, 2000.

Dukkipati, R.V., *Vibration Analysis,* Narosa Publishing House, New Delhi, India, 2005.

Fertis, D.G., *Mechanical and Structural Vibrations,* Wiley, New York, NY, 1995.

Garg, V.K., and Dukkipati, R.V., *Dynamics of Railway Vehicle Systems,* Academic Press, New York, NY, 1984.

Garg, V.K., and Dukkipati, R.V., *Dynamics of Railway Vehicle Systems,* Academic Press, New York, NY, 1984.

Genta, G., *Vibration of Structures and Machines,* Springer-Verlag, New York, NY, 1992.

Ginsberg, J.H., *Mechanical and Structural Vibrations,* Wiley, New York, NY, 2001.

Gorman, D.J., *Free Vibration Analysis of Beams and Shafts,* Wiley, New York, NY, 1975.

Gorman, D.J., *Free Vibration Analysis of Rectangular Plates,* Elsevier, 1982.

Gough, W., Richards, J.P.G., and Williams, R.P., *Vibrations and Waves,* Wiley, New York, NY, 1983.

Gross, E.E., *Measurement of Vibration,* General Radio, 1955.

Grover, G.K., *Mechanical Vibration,* Nem Chand and Bros. Roorkee, 1972.

Haberman, C.M., *Vibration Analysis,* Merril, Columbus, OH, 1968.

Hansen, H.M., and Chenea, P.F., *Mechanics of Vibration,* Wiley, New York, NY, 1952.

Harris, C.M., Crede, C.E., *Shock and Vibration Handbook,* 4th ed., McGraw Hill, New York, NY, 199

Hatter, D.H., *Matrix Computer Methods of Vibration Analysis,* Wiley, New York, NY, 1973.

Hayashi, C., *Nonlinear Oscillations in Physical Systems,* McGraw Hill. New York, NY, 1964.

Hurty, W.C., and Rubenstein, M.F., *Dynamics of Structures,* Prentice Hall, NJ, 1964.

Huston, R., and Josephs, H., *Dynamics of Mechanical Systems,* CRC Press, Boca Raton, FL, 2002.

Inman, D.J., *Vibration with Control Measurement and Stability,* Prentice Hall, Englewood cliffs, NJ, 1989.

Jackson, C., *The Practical Vibration Primer,* Gulf Publishing, Houston, TX, 1979.

Jacobsen, L.S., and Ayre, R.S., *Engineering Vibrations,* McGraw Hill, New York, 1958.

James, M.L., Smith, G.M., Wolford, J.C., and Whaley, P.W., *Vibration of Mechanical and Structural Systems,* Harper and Row, 1989.

Jones, D.S., *Electrical and Mechanical Oscillations,* Routledge and Kegan, London, 1961.

Karnopp, D.C., Margolis, D.L., and Rosenberg, R.C., *System Dynamics*, 3rd ed., Wiley Inter Science, New York NY, 2000.

Kelly, S.G., *Fundamentals of Mechanical Vibration*, McGraw Hill, New York, NY, 1993.

Kelly, S.G., *Theory and Problems of Mechanical Vibrations*, Schaum's Outline Series, McGraw Hill, New York, NY, 1996.

Kimball, A.L., *Vibration Prevention in Engineering*, Wiley, New York, NY, 1932.

Lalanne, M., Berthier, P., and Der Hagopian, J., *Mechanical Vibrations for Engineers,* Wiley, New York, NY, 1983.

Lancaster, P., *Lambda-Matrices and Vibrating Systems*, Pergamon, 1966.

Loewy, R.G., and Piarulli, V.J., *Dynamics of Rotating Shafts*, Naval Publication, 1969.

Manley, R.G., *Fundamentals of Vibration Study*, Wiley, New York, NY, 1942.

Marguerre, K., and Wolfel, H., *Mechanics of Vibration,* Sitjthoff and Noordhoff, 1979.

Mclachlan, N.W., *Theory of Vibration*, Dover publications, 1951.

Meirovitch, L., *Analytical Methods in Vibrations*, Macmillan, New York, NY, 1967.

Meirovitch, L., *Elements of Vibration Analysis*, 2nd ed., McGraw Hill, New York, NY, 1986.

Meirovitch, L., *Introduction to Dynamics and Control*, Wiley, New York, NY, 1985.

Meirovitch, L., *Methods of Analytical Dynamics*, McGraw Hill, New York, NY, 1970.

Meirovitch, L., *Principles and Techniques of Vibrations*, Prentice Hall, Upper Saddle River, NJ, 1997.

Minorosky, M., *Nonlinear Oscillations*, Van Nostrand, Princeton, NJ, 1962.

Moretti, P.M., *Modern Vibrations Primer*, CRC Press, Boca Raton, FL, 2002.

Morrill, B., *Mechanical Vibration*, The Ronald Press, 1937.

Morrow, C.T., *Shock and Vibration Engineering,* Wiley, New York, NY, 1963.

Morse, P.M., *Vibration and Sound*, McGraw Hill, New York, NY, 1948.

Muller, P.C., and Schiehlen, W.O., *Linear Vibrations*, Martinus Nighoff, 1985.

Myklestad, N.O., *Fundamentals of Vibration Analysis,* McGraw Hill, New York, NY, 1956.

Nakra, B.C., Yadava, G.S., and Thurestadt, L., *Vibration Measurement and Analysis,* NPC, New Delhi, India, 1989.

Nashif, A.D., Jones, D.I.G., and Henderson, J.P., *Vibration Damping,* Wiley, New York, NY, 1985.

Nayfeh, A.H., and Mook, D.T., *Nonlinear Oscillations*, Wiley, New York, NY, 1979.

Newland, D.E., *An Introduction to Random Vibrations and Spectral Analysis*, 2nd ed., Longman, 1984.

Newland, D.E., *Mechanical Vibration Analysis and Computation,* Longman, 1989.

Newland, D.E., *Random Vibrations and Spectral Analysis*, 2nd ed., Longman, London, 1984.

Nigam, N.C., *Introduction to Random Vibrations*, MIT Press, 1983.

Norton, M.P., *Fundamentals of Noise and Vibration Analysis for Engineers*, Cambridge University Press, Cambridge, 1989.

Pain, H.J., *The Physics of Vibrations and Waves*, Wiley, New York, NY 1983.

Pippard, A.B., *The Physics of Vibration*, Cambridge University Press, Cambridge, 1978.

Piszek, K., and Niziol, J., *Random Vibrations of Mechanical Systems*, Ellis Horwood, 1986.

Prentis, J.M., and Leckie, F.A., *Mechanical Vibrations: An Introduction to Matrix Methods,* Longman, 1963.

Ramamurti, V., *Mechanical Vibration Practice With Basic Theory*, CRC Press, Boca Raton, FL, 2000.

Rao, J.S., *Advanced Theory of Vibration*, Wiley, New York, NY, 1991.

Rao, J.S., and Dukkipati, R.V., *Mechanism and Machine Theory, 2nd ed.,* Wiley Eastern, New Delhi, India, 1992.

Rao, J.S., and Gupta, K., *Introductory Course on Theory and Practice of Mechanical Vibrations,* Wiley Eastern, New Delhi, India, 1984.

Rao, S.S., *Mechanical Vibrations*, 3rd ed., Addison Wesley, Reading, MA, 1995.

Rocard, V., *General Dynamics of Vibrations*, Unger, New York, NY, 1960.

Seto, W.W., *Theory and Problems of Mechanical Vibrations*, Schaum series, McGraw Hill, New York, NY, 1964.

Shabana, A.A., *Theory of Vibration: An Introduction*, Springer-Verlag, New York, NY, 1991.

Shabana, A.A., *Theory of Vibration: Discrete and Continuous Systems*, Springer, New York, NY, 1991.

Smith, J.D., *Vibration Measurement and Analysis,* Butterworths, 1989.

Snowdon, J.C., *Vibration and Shock in Damped Mechanical Systems,* Wiley, New York, NY, 1968.

Srinivasan, P., *Mechanical Vibration Analysis*, Tata McGraw Hill, New Delhi, India, 1982.

Steidel, R.F., *An Introduction to Mechanical Vibrations*, 3rd ed., Wiley, New York, NY, 1981.

Stoker, J.J., *Nonlinear Vibrations*, Inter science, New York, NY, 1950.

Thompson, J.M.T., and Stewart, H.B., *Nonlinear Dynamics and Chaos*, Wiley, New York NY, 1986.

Thomson, W.T., and Dahleh, M.D., *Theory of Vibrations with Applications,* 5th ed., Prentice Hall, Englewood Cliffs, NJ, 199.

Thornson, D.L., *Mechanics Applied to Vibrations and Balancing*, Wiley, New York, NY, 1940.

Timoshenko, S., Young, D.H., and Weaver, W., *Vibration Problems in Engineering*, 5th ed., Wiley, New York, NY 1990.

Timoshenko, S.P., and Young, D.H., *Advanced Dynamics*, McGraw Hill, New York, NY, 1948.

Timoshenko, S.P., *Vibrations in Engineering*, D. Van Nostrand, New York, NY, 1955.

Tong, K.N., *Theory of Mechanical Vibration*, Wiley, New York, NY, 1960.

Tse, F.S., Morse, I.E., and Hinkle, R.T., *Mechanical Vibrations,* Allyn and Bacon, Boston, MA, 1963.

Tuplin, W.A., *Torsional Vibration*, Wiley, New York, NY, 1934.

Van Santen, G.W., *Mechanical Vibration*, Macmillan, New York, NY, 1998.

Vernon, J.B., *Linear Vibration Theory*, Wiley, New York, NY, 1967.

Vierck, R.K., *Vibration Analysis,* 2nd ed., Harper & Row, New York, NY, 1979.

Volterra, E., and Zachmanoglon, E.C., *Dynamics of Vibrations*, Merrill, 1965.

Wallace, R.H., *Understanding and Measuring Vibrations*, Springer, New York, NY, 1970.

Walshaw, A.C., *Mechanical Vibrations with Applications*, Ellis Harwood, 1984.

Weaver, W., Timoshenko, S.P., and Young, D.H., *Vibration Problems in Engineering*, 5th ed., Wiley, New York, NY, 1990.

Wilson, W.K., *Practical Solution of Torsional Vibration Problems*, Vol.1, Wiley, New York, NY, 1949

Wilson, W.K., *Practical Solution of Torsional Vibration Problems*, Vol.2, Wiley, New York, NY, 1949

Wowk, V., *Machinery Vibration: Measurement and Analysis*, McGraw Hill, New York, NY, 1991.

NUMERICAL METHODS

Akai, T.J., *Applied Numerical Methods for Engineers*, Wiley, New York, NY, 1993.

Ali, R., "Finite difference methods in vibration analysis", *Shock and Vibration Digest*, Vol.15, March 1983, pp.3-7.

Atkinson, K.E., *An Introduction to Numerical Analysis*, 2nd ed., Wiley, New York, NY, 1993.

Atkinson, L.V., and Harley, P.J., *Introduction to Numerical Methods with PASCAL*, Addison Wesley, Reading, MA, 1984.

Ayyub, B.M., and McCuen, R.H., *Numerical Methods for Engineers*, Prentice Hall, Upper Saddle River, New Jersey, NJ, 1996.

Bathe, K.J., and Wilson, E.L., *Numerical Methods in Finite Element Analysis*, Prentice Hall, Englewood Cliffs, NJ, 1976.

Belytschko, T., "Explicit Time Integration of Structure-Mechanical Systems", in J. Donea (Ed.), *Advanced Structural Dynamics*, Applied Science Publishers, London, England, 1980, pp.97-122.

Belytschko, T., and Mullen, R., "Stability of Explicit-Implicit Mesh Partitions in Time Integration", *International Journal for Numerical Methods in Engineering*, Vol.12, 1975, pp.1575-1586.

Belytschko, T., Schoeberle. D.F., "On the Unconditional Stability of An Implicit Algorithm for Nonlinear Structural Dynamics", *Journal of Applied Mechanics*, Vol.42, 1975, pp.865-869.

Belytschko. T., Holmes, N., and Mullen, R., "Explicit Integration Stability, Solution Properties, Cost", *Finite-Element Analysis of Transient Nonlinear Structural Behavior,* ASME, AMD Vol.14, 1975.

Bhat, R.B., and Dukkipati, R.V., *Advanced Dynamics,* Narosa Publishing House, New Delhi, India, 2001.

Brice, C., Luther, H.A and Wilkes, J. O., *Applied Numerical Methods*, New York, NY, 1969.

Chapra, S.C., *Numerical Methods for Engineers with Software and Programming Applications*, 4th ed., McGraw Hill, New York, NY, 2002.

Clough, R.W., and Penzien, J., *Dynamics of Structures*, McGraw Hill, New York, NY, 1975.

Conte, S.D., and DeBoor, C.W., *Elementary Numerical Analysis: An Algorithm Approach*, 2nd ed., McGraw Hill, New York, NY, 1972.

Cornwell, R.E., Craig, R.R. Jr., and Johnson, C.P., "On the Application of the Mode-Acceleration Method to Structural Engineering Problems", *Earthquake Engineering and Structural Dynamics*, Vol. 11, 1983, pp. 679-688.

Dukkipati, R.V., Ananda Rao, M., and Bhat, R.B., *Computer Aided Analysis and Design of Machine Elements,* Narosa Publishing House, New Delhi, India, 2000.

Dukkipati, R.V., and Amyot, J.R., *Computer Aided Simulation in Railway Vehicle Dynamics*, Marcel-Dekker, New York, NY, 1988.

Dukkipati, R.V., *Vehicle Dynamics,* Narosa Publishing House, New Delhi, India, 2000.

Epperson, J.F., *An Introduction to Numerical Methods and Analysis*, Wiley, New York, NY, 2001.

Fallow, S.J., "A Computer Program to Find Analytical Solutions of Second Order Linear Differential Equations", *International Journal for Numerical Methods in Engineering*, Vol.6, 1973, pp. 603-606.

Fausett, L.V., *Applied Numerical Analysis using MATLAB*, Prentice Hall, Upper Saddle River, New Jersey, NJ, 1999.

Fausett, L.V., *Numerical Methods using MATHCAD*, Prentice Hall, Upper Saddle River, New Jersey, NJ, 2002.

Ferziger, J.H., *Numerical Methods for Engineering Applications*, 2nd ed., Wiley, New York, NY, 1998.

Forbear, C. E., *Introduction to Numerical Analysis*, Addison Wesley, Reading, MA, 1969.

Garg, V.K., and Dukkipati, R.V., *Dynamics of Railway Vehicle Systems*, Academic Press, New York, NY, 1984.

Gerald, C.F., and Wheatley, P.O., *Applied Numerical Analysis*, 3rd ed., Addison Wesley, Reading, MA, 1984.

Goudreau, G.L., and Taylor, R.L., "Evaluation of Numerical Integration Methods in Elastodynamics", *Computational Methods in Applied Mechanics and Engineering*, Vol. 2, 1973, pp. 69-97.

Hanselman, D., and Littlefield, B.R., *Mastering MATLAB 6*, Prentice Hall, Upper Saddle River, New Jersey, NJ, 2001.

Hildebrand, F.B., *Introduction to Numerical Analysis*, McGraw-Hill, New York, NY, 1956.

Hojjat, A., Gere, J.M., and Weaver, W., "Algorithm For Nonlinear Structural Dynamics", *Journal of the Structural Division*, ASCE, Feb.1978, pp. 263-279.

Houbolt, J.C., "A Recurrence Matrix Solution for the Dynamic Response of Elastic Aircraft", *Journal of Aeronautical Sciences*, Vol.17, 1950, pp. 540-550, 594.

Huges, T.J.R., "A Note on the Stability of Newmark's Algorithm in Nonlinear Structural Dynamics", *International Journal for Numerical Methods in Engineering*, Vol.11, 1976, pp. 383-386.

Hurty, W.C., and Rubinstein, M.F., *Dynamics of Structures*, Prentice Hall, Englewood Cliffs, NJ, 1970.

Jennings, A., and Orr, D.R.L., "Application of the Simultaneous Iteration Method to Undamped Vibration Problems", *International Journal for Numerical Methods in Engineering*, Vol.3, 1971, pp.13-24.

Key, S.W., "Transient Response by Time Integration: Review of Implicit and Explicit Operators", in J. Donea (Ed.), *Advanced Structural Dynamics*, Applied Science Publishers, London, England, 1980.

Krieg, R.D., "Unconditional Stability in Numerical Time Integration Methods", *Journal of Applied Mechanics*, Vol.40, 1973, pp.417-421.

Lambert, J.D., *Numerical Methods for Ordinary Differential Equations—The Initial Value Problems,* Wiley, New York, NY, 1991.

Lau, P.C.M., "Finite Difference Approximation for Ordinary Derivatives", *International Journal for Numerical Methods in Engineering,* Vol.17, 1981, pp.663-678.

Leech, J.W., Hsu, P.T., Mack, E.W., "Stability of A Finite-Difference Method for Solving Matrix Equations", *AIAA Journal*, Vol.3, 1965, pp. 2172-2173.

Levy, S., and Kroll W.D., "Errors Introduced by Finite Space and Time Increments in Dynamics Response Computation", *Proceedings of the First U.S. National Congress of Applied Mechanics*, 1951, pp.1-8.

Levy, S., and Wilkinson, J.P.D., *The Component Element Method in Dynamics with Application to Earthquake Engineering*, McGraw Hill, New York, NY, 1976.

Lindfield, G., and Penny, J., *Numerical Methods using MATLAB*, 2nd ed., Prentice Hall, Upper Saddle River, New Jersey, NJ, 2000.

Magrab, E.B., *An Engineers Guide to MATLAB*, Prentice Hall, Upper Saddle River, New Jersey, NJ, 2001.

Mathews, J.H., and Fink, K., *Numerical Methods using MATLAB*, 3rd ed., Prentice Hall, Upper Saddle River, New Jersey, NJ, 1999.

McNamara, J.F., "Solution Schemes for Problems of Nonlinear Structural Dynamics", *Journal of Pressure Vessel Technology,* ASME, May 1974, pp.96-102.

Nakamara, S., *Numerical Analysis and Graphic Visualization with MATLAB,* 2nd ed., Prentice Hall, Upper Saddle River, New Jersey, NJ, 2002.

Nakamura, S., *Computational Methods in Engineering and Science,* Wiley, New York, NY, 1977.

Newmark, N.M., "A Method of Computation for Structural Dynamics", *ASCE Journal of Engineering Mechanics Division,* Vol. 85, 1959, pp. 67-94.

Park, K.C., "An improved Stiffly Method for Direct Integration of Non-Linear Structural Dynamics Equations", *Journal of the Applied Mechanics,* ASME, June 1975, pp.464-470.

Penman, E.D., "A Numerical Method for Coupled Differential Equations", *International Journal for Numerical Methods in Engineering,* Vol. 41972, pp. 587-596.

Rao, S.S., *Applied Numerical Methods for Engineers and Scientists,* Prentice Hall, Upper Saddle River, New Jersey, NJ, 2002.

Reali, M., Rangogni, R., and Pennati, V., "Compact Analytic Expressions of Two-Dimensional Finite Difference Forms", *International Journal for Numerical Methods in Engineering,* Vol. 20, 1984, pp.121-130.

Recktenwald, G.W., *Introduction to Numerical Methods and MATLAB—Implementation and Applications,* Prentice Hall, Upper Saddle River, New Jersey, NJ, 2001.

Romanelli, M.J., "Runge-Kutta Method for the Solution of Ordinary Differential Equations", in *Mathematical Methods for Digital Computers,* A. Ralston and H.S. Wilf (eds.), Wiley, New York, NY, 1965.

Tillerson, J.R., Stricklin, J.A., and Haisler, W.E., "Numerical Methods for the Solution of Nonlinear Problems in Structural Analysis", *ASME Winter Annual Meeting,* Detroit, MI, Nov. 11-15, 1973.

Timoshenko, S.P., Young, D.H., and Weaver, W. Jr., *Vibration Problems in Engineering,* 4th ed., Wiley, New York, 1974.

Wah, T., and Colcote, L.R., *Structural Analysis by Finite Difference Calculus,* Van Nostrand Reinhold, New York, NY, 1970.

Wang, P.C., *Numerical and Matrix Methods in Structural Mechanics,* Wiley, New York, NY, 1966.

Wilson, E.L., Farhoomand, I., and Bathe, K.J., "Nonlinear Dynamic Analysis of Complex Structures", *International Journal of Earthquake Engineering and Structural Dynamics,* Vol. 1, 1973, pp. 241-252.

MATLAB

Chapman, S.J., *MATLAB Programming for Engineers,* 2nd ed., Brooks/Cole, Thomson Learning, Pacific Grove, CA, 2002.

Dabney, J.B., and Harman, T.L., *Mastering SIMULINK 4,* Prentice Hall, Upper Saddle River, NJ, 2001.

Djaferis, T.E., *Automatic Control- The Power of Feedback using MATLAB*, Brooks/Cole, Thomson Learning, Pacific Grove, CA, 2000.

Dukkipati, R.V., *Solving Engineering Mechanics Problems with MATLAB*, New Age International (P) Ltd., New Delhi, India, ISBN: 81-224-1809-0, 2007.

Dukkipati, R.V., *MATLAB for Engineers*, New Age International (P) Ltd., New Delhi, India, ISBN: 81-224-1809-0, 2007.

Dukkipati, R.V., *Analysis and Design of Control Systems using MATLAB*, New Age International (P) Ltd., New Delhi, India, ISBN: 81-224-1809-0, 2006.

Dukkipati, R.V., *Solving Engineering System Dynamics Problems with MATLAB*, New Age International (P) Ltd., New Delhi, India, 2007.

Etter, D.M., *Engineering Problem Solving with MATLAB*, Prentice-Hall, Englewood Cliffs, NJ, 1993.

Gardner, J.F., *Simulation of Machines using MATLAB and SIMULINK*, Brooks/Cole, Thomson Learning, Pacific Grove, CA, 2001.

Harper, B. D., *Solving Dynamics Problems in MATLAB*, 5th ed, Wiley, New York, 2002.

Harper, B. D., *Solving Statics Problems in MATLAB*, 5th ed, Wiley, New York, 2002.

Herniter, M.E., *Programming in MATLAB*, Brooks/Cole, Pacific Grove, CA, 2001.

Karris, S.T., *Signals and Systems with MATLAB Applications*, Orchard Publications, Fremont, CA, 2001.

Leonard, N.E., and Levine, W.S., *Using MATLAB to Analyze and Design Control Systems*, Addison-Wesley, Redwood City, CA, 1995.

Lyshevski, S.E., *Engineering and Scientific Computations Using MATLAB*, Wiley, New York, 2003.

Moler, C., *The Student Edition of MATLAB for MS-DOS Personal Computers with 3-1/ 2" Disks*, MATLAB Curriculum Series, The MathWorks, Inc., 2002.

Ogata, K., *Designing Linear Control Systems with MATLAB*, Prentice Hall, Upper Saddle River, NJ, 1994.

Ogata, K., *Solving Control Engineering Problems with MATLAB*, Prentice Hall, Upper Saddle River, NJ, 1994.

Pratap, Rudra., *Getting Started with MATLAB- A Quick Introduction for Scientists and Engineers*, Oxford University Press, New York, NY, 2002.

Saadat, Hadi., *Computational Aids in Control Systems using MATLAB*, McGraw Hill, New York, NY, 1993.

Sigman,K., and Davis, T.A., *MATLAB Primer*, 6th ed, Chapman& Hall/CRCPress, Boca Raton, FL, 2002.

The MathWorks, Inc., *SIMULINK, Version 3*, The MathWorks, Inc., Natick, MA, 1999.

The MathWorks, Inc., *MATLAB: Application Program Interface Reference Version 6*, The MathWorks, Inc., Natick, 2000.

The MathWorks, Inc., *MATLAB: Control System Toolbox User's Guide, Version 4*, The MathWorks, Inc., Natick, 1992-1998.

The MathWorks, Inc., *MATLAB: Creating Graphical User Interfaces, Version 1*, The MathWorks, Inc., Natick, 2000.

The MathWorks, Inc., *MATLAB: Function Reference,* The MathWorks, Inc., Natick, 2000.

The MathWorks, Inc., *MATLAB: Release Notes for Release 12*, The MathWorks, Inc., Natick, 2000.

The MathWorks, Inc., *MATLAB: Symbolic Math Toolbox User's Guide, Version 2*, The MathWorks, Inc., Natick, 1993-1997.

The MathWorks, Inc., MATLAB: *Using MATLAB Graphics, Version 6*, The MathWorks, Inc., Natick, 2000.

The MathWorks, Inc.,, *MATLAB: Using MATLAB, Version 6*, The MathWorks, Inc., Natick, 2000.

JOURNALS

AIAA Journal

Applied Mechanics Reviews

ASCE Journal of Applied Mechanics

ASME Journal of Applied Mechanics

ASME Journal of Vibration and Acoustics

Bulletin of the Japan Society of Solids and Structures

Communications in Numerical Methods in Engineering

Earthquake Engineering and Structural Dynamics

International Journal for Numerical Methods in Engineering

International Journal for Numerical Methods in Engineering

International Journal of Analytical and Experimental Modal Analysis

International Journal of Vehicle Design

Journal of Mechanical Systems and Signals, Academic Press, New York, NY, USA.

Journal of Sound and Vibration, Academic Press, New York, NY, USA

Journal of the Acoustical Society of America

Journal of Vibration and Acoustics, American Society of Mechanical Engineers, New York, NY, USA.

JSME International Journal Series III - Vibration Control Engineering

Noise and Vibration Worldwide

Shock and Vibration, IOS press, Amsterdam, The Netherlands

Vehicle System Dynamics

Vibrations, Mechanical Systems and Signal Processing.

PERIODICALS

Shock and Vibration Digest, Sage Science Press, Thousand Oaks, CA, USA.

Sound and Vibration, Acoustical Publications, Bay Village, Ohio, USA.